Progress in Nonlinear Differential Equations and Their Applications
Volume 44

Editor
Haim Brezis
Université Pierre et Marie Curie
Paris
and
Rutgers University
New Brunswick, N.J.

Cristian E. Gutiérrez

The Monge–Ampère Equation

Birkhäuser
Boston • Basel • Berlin

Cristian E. Gutiérrez
Department of Mathematics
Temple University
Philadelphia, PA 19122
U.S.A.

Library of Congress Cataloging-in-Publication Data

Gutiérrez, Cristian E., 1950-
    The Monge–Ampère equation / Cristian E. Gutiérrez.
        p. cm. – (Progress in nonlinear differential equations and their applications ; v. 44)
    Includes bibliographical references and index.
    ISBN 0-8176-4177-7 (alk. paper) – ISBN 3-7643-4177-7 (alk. paper)
        1. Monge–Ampère equations. I. Title. II. Series.

    QA377.G87 2001
    515'.353–dc21   2001025444

AMS Subject Classifications: 35J60, 35J65, 53A15, 52A20

Printed on acid-free paper.
©2001 Birkhäuser Boston

Birkhäuser ®

ISBN 0-8176-4177-7   SPIN 10761705
ISBN 3-7643-4177-7

Reformatted from author's files in LATEX 2ε by TEXniques, Inc., Cambridge, MA.
Printed and bound by Hamilton Printing Company, Rensselaer, NY.
Printed in the United States of America.

9 8 7 6 5 4 3 2 1

To Graciela, Romina, and Daniel

Contents

Preface

In recent years, the study of the Monge–Ampère equation has received considerable attention and there have been many important advances. As a consequence there is nowadays much interest in this equation and its applications. This volume tries to reflect these advances in an essentially self-contained systematic exposition of the theory of weak solutions, including recent regularity results by L. A. Caffarelli. The theory has a geometric flavor and uses some techniques from harmonic analysis such us covering lemmas and set decompositions. An overview of the contents of the book is as follows.

We shall be concerned with the Monge–Ampère equation, which for a smooth function u, is given by

$$\det D^2 u = f. \tag{0.0.1}$$

There is a notion of generalized or weak solution to (0.0.1): for u convex in a domain Ω, one can define a measure Mu in Ω such that if u is smooth, then Mu has density $\det D^2 u$. Therefore u is a generalized solution of (0.0.1) if $Mu = f$. The notion of generalized solution is based on the notion of normal mapping, and in Chapter 1 we begin with these two concepts, introduced by A. D. Aleksandrov, and we describe their basic properties. The notion of viscosity solution is also considered and compared with that of generalized solution. We also introduce several maximum principles that are fundamental in the study of the Monge–Ampère operator. The Dirichlet problem for Monge–Ampère is then solved in the class of generalized solutions in Sections 1.5 and 1.6. Chapter 1 concludes with the concept of ellipsoid of minimum volume which is of particular importance in developing the theory of cross-sections in Chapter 3.

In Chapter 2, we present the Harnack inequality of Krylov–Safonov for non-divergence elliptic operators in view of some ideas used to study the linearized Monge–Ampère equation. This illustrates these ideas in a case that is simpler than that of the linearized Monge–Ampère operator.

Chapter 3 presents the theory of cross-sections of weak solutions to the Monge–Ampère equation and we prove several geometric properties that are needed in the subsequent chapters. The cross-sections of u are the level sets of the convex function u minus a supporting hyperplane. Of special importance is the doubling condition (3.1.1) for the measure Mu that permits us, from the characterization given in Theorem 3.3.5, to determine invariance properties for the shapes of cross-sections that are valid under appropriate normalizations using ellipsoids of minimum volume. A typical situation is when the measure Mu satisfies

$$\lambda \, |E| \leq Mu(E) \leq \Lambda \, |E|, \qquad (0.0.2)$$

for some positive constants λ, Λ and for all Borel subsets E of the convex domain Ω. The inequalities (0.0.2) resemble the uniform ellipticity condition for linear operators. The results proved in this chapter permit us to work with the cross-sections as if they were Euclidean balls and to establish the covering lemmas needed later for the regularity theory in Chapters 4–6.

Chapter 4 concerns an application of the properties of the sections: a result of Jörgens–Calabi–Pogorelov–Cheng and Yau about the characterization of global solutions of $Mu = 1$.

Chapter 5 contains Caffarelli's $C^{1,\alpha}$ estimates for weak solutions. A fundamental geometric result is Theorem 5.2.1 about the extremal points of the set where a solution u equals a supporting hyperplane.

Finally, in Chapter 6 we present the $W^{2,p}$ estimates for the Monge–Ampère equation recently developed by Caffarelli and extending classical estimates of Pogorelov. The main result here is Theorem 6.4.2.

We have included bibliographical notes at the end of each chapter.

Acknowledgments

It is a pleasure to thank all the people who assisted me during the preparation of this book. I am particularly indebted to L. A. Caffarelli for inspiration, many discussions, and for his collaboration. I am very grateful to Qingbo Huang for innumerable enlightening discussions on most topics in this book, for many suggestions and corrections, and for his collaboration. I am also very grateful to several friends and students for carefully reading various chapters of the manuscript: Shif Berhanu, Giuseppe Di Fazio, David Hartenstine, and Federico Tournier. They have made many helpful comments, suggestions and corrections that improved the presentation. I would especially like to thank L. C. Evans for his encouragement and suggestions.

This book encompasses the contents of a graduate course at Temple University, and some chapters have been used in short courses at the Università di Bologna, Universidad de Buenos Aires and Universidad Autónoma de Madrid. I would like to thank these institutions and all my friends there for the kind hospitality and support.

The research connected with the results in this volume was supported in part by the National Science Foundation and I wish to thank this institution for its support.

Cristian E. Gutiérrez
Cherry Hill, New Jersey
September 2000

Notation

Du denotes the gradient of the function u.

$D^2u(x)$ denotes the Hessian of the function u, i.e., $D^2u(x) = \left(\dfrac{\partial^2 u(x)}{\partial x_i \partial x_j}\right)$,

$1 \le i, j \le n$.

$\Omega \subset \mathbb{R}^n$, $u : \Omega \to \mathbb{R}$ is convex if for all $0 \le t \le 1$ and any $x, y \in \Omega$ such that $t x + (1 - t) y \in \Omega$ we have

$$u(t x + (1 - t) y) \le t u(x) + (1 - t)u(y).$$

Given a set E, $\chi_E(x)$ denotes the characteristic function of E.

$|E|$ denotes the Lebesgue measure of the set E.

$B_R(x)$ denotes the Euclidean ball centered at x with radius R.

ω_n denotes the measure of the unit ball in $\mathbb{R}^n$.

$C(\Omega)$ denotes the class of real-valued functions that are continuous in Ω.

Given a positive integer k, $C^k(\Omega)$ denotes the class of real-valued functions that are continuously differentiable in Ω up to order k.

If E_k is a sequence of sets, then

$$E^* = \limsup_{n \to \infty} E_n = \cap_{n=1}^{\infty} \cup_{k=n}^{\infty} E_k; \qquad E_* = \liminf_{n \to \infty} E_n = \cup_{n=1}^{\infty} \cap_{k=n}^{\infty} E_k;$$

$$\chi_{E^*}(x) = \limsup_{n \to \infty} \chi_{E_n}(x); \qquad \chi_{E_*}(x) = \liminf_{n \to \infty} \chi_{E_n}(x).$$

The real-valued function u is harmonic in the open set $\Omega \subset \mathbb{R}^n$ if $u \in C^2(\Omega)$ and $\Delta u(x) = \sum_{i=1}^{n} \dfrac{\partial^2 u(x)}{\partial x_i^2} = 0$ in Ω.

If $\Omega \subset \mathbb{R}^n$ is a bounded and measurable set, the center of mass or baricenter of Ω is the point x^* defined by

$$x^* = \frac{1}{|\Omega|} \int_{\Omega} x \, dx.$$

If $A \subset B \subset \mathbb{R}^n$ and $\bar{A} \subset B$, then we write $A \Subset B$.

If $a, b \in \mathbb{R}$, then $a \vee b = \max\{a, b\}$.

If E is a set, then $\mathcal{P}(E)$ denotes the class of all subsets of E.

If $Q \subset \mathbb{R}^n$ is a cube and $\alpha > 0$, then αQ denotes the cube concentric with Q but with edge length equals α times the edge length of Q.

Chapter 1

Generalized Solutions to Monge–Ampère Equations

1.1 The normal mapping

Let Ω be an open subset of $\mathbb{R}^n$ and $u : \Omega \to \mathbb{R}$. Given $x_0 \in \Omega$, *a supporting hyperplane to the function u at the point $(x_0, u(x_0))$ is an affine function $\ell(x) = u(x_0) + p \cdot (x - x_0)$ such that $u(x) \geq \ell(x)$ for all $x \in \Omega$.*

Definition 1.1.1 *The normal mapping of u, or subdifferential of u, is the set-valued function $\partial u : \Omega \to \mathcal{P}(\mathbb{R}^n)$ defined by*

$$\partial u(x_0) = \{p : u(x) \geq u(x_0) + p \cdot (x - x_0), \quad \text{for all } x \in \Omega\}.$$

Given $E \subset \Omega$, we define $\partial u(E) = \bigcup_{x \in E} \partial u(x)$.

The set $\partial u(x_0)$ may be empty. Let $S = \{x \in \Omega : \partial u(x) \neq \emptyset\}$. If $u \in C^1(\Omega)$ and $x \in S$, then $\partial u(x) = Du(x)$, the gradient of u at x, which means that when u is differentiable the normal mapping is basically the gradient. If $u \in C^2(\Omega)$ and $x \in S$, then the Hessian of u is nonnegative definite, that is $D^2 u(x) \geq 0$. This means that if u is C^2, then S is the set where the graph of u is concave up. Indeed, by Taylor's Theorem $u(x + h) = u(x) + Du(x) \cdot h + \frac{1}{2}\langle D^2 u(\xi)h, h \rangle$, where ξ lies on the segment between x and $x + h$. Since $u(x + h) \geq u(x) + Du(x) \cdot h$ for all h sufficiently small, the claim follows.

Example 1.1.2 It is useful to calculate the normal mapping of the function u whose graph is a cone in $\mathbb{R}^{n+1}$. Let $\Omega = B_R(x_0)$ in $\mathbb{R}^n$, $h > 0$ and $u(x) = h \dfrac{|x - x_0|}{R}$. The graph of u, for $x \in \Omega$, is an upside-down right-cone in $\mathbb{R}^{n+1}$ with vertex at the point $(x_0, 0)$ and base on the hyperplane $x_{n+1} = h$. We shall show that

$$\partial u(x) = \begin{cases} \dfrac{h}{R} \dfrac{x - x_0}{|x - x_0|}, & \text{for } 0 < |x - x_0| < R, \\[2mm] \overline{B_{h/R}(0)}, & \text{for } x = x_0. \end{cases}$$

If $0 < |x - x_0| < R$, then the value of ∂u follows by calculating the gradient. By the definition of normal mapping, $p \in \partial u(x_0)$ if and only if $\dfrac{h}{R}|x - x_0| \geq p \cdot (x - x_0)$ for all $x \in B_R(x_0)$. If $p \neq 0$ and we pick $x = x_0 + R\dfrac{p}{|p|}$, then $|p| \leq \dfrac{h}{R}$. It is clear that $|p| \leq \dfrac{h}{R}$ implies $p \in \partial u(x_0)$.

1.1.1 Properties of the normal mapping

Lemma 1.1.3 *If $\Omega \subset \mathbb{R}^n$ is open, $u \in C(\Omega)$ and $K \subset \Omega$ is compact, then $\partial u(K)$ is compact.*

Proof. Let $\{p_k\} \subset \partial u(K)$ be a sequence. We claim that p_k is bounded. For each k there exists $x_k \in K$ such that $p_k \in \partial u(x_k)$, that is $u(x) \geq u(x_k) + p_k \cdot (x - x_k)$ for all $x \in \Omega$. Since K is compact, $K_\delta = \{x : \text{dist}(x, K) \leq \delta\}$ is compact and contained in Ω for all δ sufficiently small, and we may assume by passing if necessary through a subsequence that $x_k \to x_0$. Then $x_k + \delta w \in K_\delta$, and $u(x_k + \delta w) \geq u(x_k) + \delta p_k \cdot w$ for all $|w| = 1$ and for all k. If $p_k \neq 0$ and $w = \dfrac{p_k}{|p_k|}$ then we get $\max_{K_\delta} u(x) \geq \min_K u(x) + \delta|p_k|$, for all k. Since u is locally bounded, the claim is proved. Hence there exists a convergent subsequence $p_{k_m} \to p_0$. We claim that $p_0 \in \partial u(K)$. We shall prove that $p_0 \in \partial u(x_0)$. We have $u(x) \geq u(x_{k_m}) + p_{k_m} \cdot (x - x_{k_m})$ for all $x \in \Omega$ and, since u is continuous, by letting $m \to \infty$ we obtain $u(x) \geq u(x_0) + p_0 \cdot (x - x_0)$ for all $x \in \Omega$. This completes the proof of the lemma. ∎

Remark 1.1.4 We note that the proof above shows that if u is only locally bounded in Ω, then $\partial u(E)$ is bounded whenever E is bounded with $\overline{E} \subset \Omega$.

Remark 1.1.5 We note that given $x_0 \in \Omega$, the set $\partial u(x_0)$ is convex. However, if K is convex and $K \subset \Omega$, then the set $\partial u(K)$ is not necessarily convex. An example is given by $u(x) = e^{|x|^2}$ and $K = \{x \in \mathbb{R}^n : |x_i| \leq 1, \quad i = 1, \ldots, n\}$. The set $\partial u(K)$ is a star-shaped symmetric set around the origin that is not convex, see Figure 1.1.

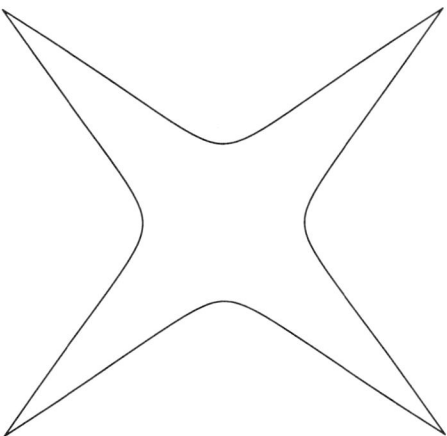

Figure 1.1. $\partial u(K)$

Lemma 1.1.6 *If u is a convex function in Ω and $K \subset \Omega$ is compact, then u is uniformly Lipschitz in K, that is, there exists a constant $C = C(u, K)$ such that $|u(x) - u(y)| \leq C|x - y|$ for all $x, y \in K$.*

Proof.   Since u is convex, u has a supporting hyperplane at any $x \in \Omega$. Let $C = \sup\{|p| : p \in \partial u(K)\}$. By Lemma 1.1.3, $C < \infty$. If $x \in K$, then $u(y) \geq u(x) + p \cdot (y - x)$ for $p \in \partial u(x)$ and for all $y \in \Omega$. In particular, if $y \in K$, then $u(y) - u(x) \geq -|p||y - x|$. By reversing the roles of x and y we get the lemma. ∎

Lemma 1.1.7 *If Ω is open and u is Lipschitz continuous in Ω, then u is differentiable a.e. in Ω.*

Proof.   See [EG92, p. 81]. ∎

Lemma 1.1.8 *If u is convex or concave in Ω, then u is differentiable a.e. in Ω.*

Proof.   Follows immediately from Lemmas 1.1.6 and 1.1.7. ∎

Remark 1.1.9 A deep result of Busemann–Feller–Aleksandrov establishes that any convex function in Ω has second order derivatives a.e., see [EG92, p. 242] and [Sch93, pp. 31–32].

Definition 1.1.10 *The Legendre transform of the function* $u : \Omega \to \mathbb{R}$ *is the function* $u^* : \mathbb{R}^n \to \mathbb{R}$ *defined by*

$$u^*(p) = \sup_{x \in \Omega} (x \cdot p - u(x)).$$

Remark 1.1.11 If Ω is bounded and u is bounded in Ω, then u^* is finite. Also, u^* is convex in $\mathbb{R}^n$.

Lemma 1.1.12 *If Ω is open and u is a continuous function in Ω, then the set of points in $\mathbb{R}^n$ that belong to the image by the normal mapping of more than one point of Ω has Lebesgue measure zero. That is, the set*

$$S = \{p \in \mathbb{R}^n : there\ exist\ x, y \in \Omega,\ x \neq y\ and\ p \in \partial u(x) \cap \partial u(y)\}$$

has measure zero. This also means that the set of supporting hyperplanes that touch the graph of u at more than one point has measure zero.

Proof. We may assume that Ω is bounded because otherwise we write $\Omega = \cup_k \Omega_k$, where $\Omega_k \subset \Omega_{k+1}$ are open and $\overline{\Omega_k}$ are compact. If $p \in S$, then there exist $x, y \in \Omega$, $x \neq y$ and $u(z) \geq u(x) + p \cdot (z - x), u(z) \geq u(y) + p \cdot (z - y)$ for all $z \in \Omega$. Since Ω_k increases, $x, y \in \Omega_m$ for some m and obviously the previous inequalities hold true for $z \in \Omega_m$. That is, if

$$S_m = \{p \in \mathbb{R}^n : there\ exist\ x, y \in \Omega,\ x \neq y\ and\ p \in \partial(u|\Omega_m)(x) \cap \partial(u|\Omega_m)(y)\}$$

we have $p \in S_m$, i.e., $S \subset \cup_m S_m$ and we then show that each S_m has measure zero.

Let u^* be the Legendre transform of u. By Remark 1.1.11 and Lemma 1.1.8, u^* is differentiable a.e. Let $E = \{p : u^*$ is not differentiable at $p\}$. We shall show that

$$\{p \in \mathbb{R}^n : there\ exist\ x, y \in \Omega,\ x \neq y\ and\ p \in \partial u(x) \cap \partial u(y)\} \subset E.$$

In fact, if $p \in \partial u(x_1) \cap \partial u(x_2)$ and $x_1 \neq x_2$, then $u^*(p) = x_i \cdot p - u(x_i), i = 1, 2$. Also $u^*(z) \geq x_i \cdot z - u(x_i)$ and so $u^*(z) \geq u^*(p) + x_i \cdot (z - p)$ for all $z, i = 1, 2$. Hence if u^* were differentiable at p we would have $Du^*(p) = x_i, i = 1, 2$. This completes the proof of the lemma. ∎

Theorem 1.1.13 *If Ω is open and $u \in C(\Omega)$, then the class*

$$\mathcal{S} = \{E \subset \Omega : \partial u(E)\ is\ Lebesgue\ measurable\}$$

is a Borel σ-algebra. The set function $Mu : \mathcal{S} \to \overline{\mathbb{R}}$ defined by

$$Mu(E) = |\partial u(E)| \tag{1.1.1}$$

is a measure, finite on compacts, that is called the Monge–Ampère measure associated with the function u.

Proof. By Lemma 1.1.3, the class $\mathcal{S}$ contains all compact subsets of Ω. Also, if E_m is any sequence of subsets of Ω, then $\partial u\,(\cup_m E_m) = \cup_m \partial u(E_m)$. Hence, if $E_m \in \mathcal{S}, m = 1, 2, \ldots$, then $\cup_m E_m \in \mathcal{S}$. In particular, we may write $\Omega = \cup_m K_m$ with K_m compacts and we obtain $\Omega \in \mathcal{S}$. To show that $\mathcal{S}$ is a σ-algebra it remains to show that if $E \in \mathcal{S}$, then $\Omega \setminus E \in \mathcal{S}$. We use the following formula, which is valid for any set $E \subset \Omega$:

$$\partial u(\Omega \setminus E) = (\partial u(\Omega) \setminus \partial u(E)) \cup (\partial u(\Omega \setminus E) \cap \partial u(E)). \tag{1.1.2}$$

By Lemma 1.1.12, $|\partial u(\Omega \setminus E) \cap \partial u(E)| = 0$ for any set E. Then from (1.1.2) we get $\Omega \setminus E \in \mathcal{S}$ when $E \in \mathcal{S}$.

We now show that Mu is σ-additive. Let $\{E_i\}_{i=1}^\infty$ be a sequence of disjoint sets in $\mathcal{S}$ and set $\partial u(E_i) = H_i$. We must show that

$$\left| \partial u\left(\cup_{i=1}^\infty E_i\right) \right| = \sum_{i=1}^\infty |H_i|.$$

Since $\partial u\left(\cup_{i=1}^\infty E_i\right) = \cup_{i=1}^\infty H_i$, we shall show that

$$\left| \cup_{i=1}^\infty H_i \right| = \sum_{i=1}^\infty |H_i|. \tag{1.1.3}$$

We have $E_i \cap E_j = \emptyset$ for $i \neq j$. Then by Lemma 1.1.12 $|H_i \cap H_j| = 0$ for $i \neq j$. Let us write

$$\cup_{i=1}^\infty H_i = H_1 \cup (H_2 \setminus H_1) \cup (H_3 \setminus (H_2 \cup H_1)) \cup (H_4 \setminus (H_3 \cup H_2 \cup H_1)) \cup \cdots,$$

where the sets on the right hand side are disjoint. Now

$$H_n = \left[H_n \cap (H_{n-1} \cup H_{n-2} \cup \cdots \cup H_1)\right] \cup \left[H_n \setminus (H_{n-1} \cup H_{n-2} \cup \cdots \cup H_1)\right].$$

Then by Lemma 1.1.12, $|H_n \cap (H_{n-1} \cup H_{n-2} \cup \cdots \cup H_1)| = 0$ and we obtain

$$|H_n| = |H_n \setminus (H_{n-1} \cup H_{n-2} \cup \cdots \cup H_1)|.$$

Consequently (1.1.3) follows, and the proof of the theorem is complete. ∎

Example 1.1.14 If $u \in C^2(\Omega)$ is a convex function, then the Monge–Ampère measure Mu associated with u satisfies

$$Mu(E) = \int_E \det D^2 u(x)\, dx, \tag{1.1.4}$$

for all Borel sets $E \subset \Omega$. To prove (1.1.4), we use the following result:

Theorem 1.1.15 (Sard's Theorem, see [Mil97]) *Let $\Omega \subset \mathbb{R}^n$ be an open set and $g : \Omega \to \mathbb{R}^n$ a C^1 function in Ω. If $S_0 = \{x \in \Omega : \det g'(x) = 0\}$, then $|g(S_0)| = 0$.*

We first notice that since u is convex and $C^2(\Omega)$, then Du is one-to-one on the set $A = \{x \in \Omega : D^2 u(x) > 0\}$. Indeed, let $x_1, x_2 \in A$ with $Du(x_1) = Du(x_2)$. By convexity $u(z) \geq u(x_i) + Du(x_i) \cdot (z - x_i)$ for all $z \in \Omega$, $i = 1, 2$. Hence $u(x_1) - u(x_2) = Du(x_1) \cdot (x_1 - x_2) = Du(x_2) \cdot (x_1 - x_2)$. By the Taylor formula we can write

$$u(x_1) = u(x_2) + Du(x_2) \cdot (x_1 - x_2)$$
$$+ \int_0^1 t \langle D^2 u \left(x_2 + t(x_1 - x_2) \right) (x_1 - x_2), x_1 - x_2 \rangle \, dt.$$

Therefore the integral is zero and the integrand must vanish for $0 \leq t \leq 1$. Since $x_2 \in A$, it follows that $x_2 + t(x_1 - x_2) \in A$ for t small. Therefore $x_1 = x_2$.

If $u \in C^2(\Omega)$, then $g = Du \in C^1(\Omega)$. We have $Mu(E) = |Du(E)|$ and

$$Du(E) = Du(E \cap S_0) \cup Du(E \setminus S_0).$$

Since $E \subset \mathbb{R}^n$ is a Borel set, $E \cap S_0$ and $E \setminus S_0$ are also Borel sets. Hence, by the formula of change of variables and Sard's Theorem,

$$Mu(E) = Mu(E \cap S_0) + Mu(E \setminus S_0) = \int_{E \setminus S_0} \det D^2 u(x) \, dx = \int_E \det D^2 u(x) \, dx,$$

which shows (1.1.4).

Example 1.1.16 If $u(x)$ is the cone of Example 1.1.2, then the Monge–Ampère measure associated with u is $Mu = |B_{h/R}| \delta_{x_0}$, where δ_{x_0} denotes the Dirac delta at x_0.

1.2 Generalized solutions

Definition 1.2.1 Let ν be a Borel measure defined in Ω, an open and convex subset of $\mathbb{R}^n$. The convex function $u \in C(\Omega)$ is a generalized solution, or Aleksandrov solution, to the Monge–Ampère equation

$$\det D^2 u = \nu$$

if the Monge–Ampère measure Mu associated with u defined by (1.1.1) equals ν.

The following lemma implies that the notion of generalized solution is closed under uniform limits. That is, if u_k are generalized solutions to $\det D^2 u = \nu$ in Ω and $u_k \to u$ uniformly on compact subsets of Ω, then u is also a generalized solution to $\det D^2 u = \nu$ in Ω.

Lemma 1.2.2 Let $u_k \in C(\Omega)$ be convex functions such that $u_k \to u$ uniformly on compact subsets of Ω.
    Then:

(i) If $K \subset \Omega$ is compact, then

$$\limsup_{k \to \infty} \partial u_k(K) \subset \partial u(K),$$

and by Fatou

$$\limsup_{k \to \infty} |\partial u_k(K)| \le |\partial u(K)|.$$

(ii) If K is compact and U is open such that $K \subset U \subset \overline{U} \subset \Omega$, then

$$\partial u(K) \subset \liminf_{k \to \infty} \partial u_k(U),$$

where the inequality holds for almost every point of the set on the left-hand side, and by Fatou

$$|\partial u(K)| \le \liminf_{k \to \infty} |\partial u_k(U)|.$$

Proof. (i) If $p \in \limsup_{k \to \infty} \partial u_k(K)$, then for each n there exist k_n and $x_{k_n} \in K$ such that $p \in \partial u_{k_n}(x_{k_n})$. By selecting a subsequence x_j of x_{k_n} we may assume that $x_j \to x_0 \in K$. On the other hand,

$$u_j(x) \ge u_j(x_j) + p \cdot (x - x_j), \qquad \forall x \in \Omega,$$

and by letting $j \to \infty$, by the uniform convergence of u_j on compacts we get

$$u(x) \ge u(x_0) + p \cdot (x - x_0), \qquad \forall x \in \Omega,$$

that is $p \in \partial u(x_0)$.

(ii) Let $S = \{p : p \in \partial u(x_1) \cap \partial u(x_2) \quad \text{for some } x_1, x_2 \in \Omega, x_1 \neq x_2\}$. By Lemma 1.1.12, $|S| = 0$. Let $K \subset \Omega$ be compact and consider $\partial u(K) \setminus S$. If $p \in \partial u(K) \setminus S$, then there exists a unique $x_0 \in K$ such that $p \in \partial u(x_0)$ and $p \notin \partial u(x_1)$ for all $x_1 \in \Omega, x_1 \neq x_0$. Let U be open satisfying the assumptions. If $x_1 \in \Omega$ and $x_1 \neq x_0$, then $u(x_1) > u(x_0) + p \cdot (x_1 - x_0)$. Otherwise, $u(x_1) = u(x_0) + p \cdot (x_1 - x_0)$ and since $p \in \partial u(x_0)$ we have

$$\begin{aligned}
u(x) &\ge u(x_0) + p \cdot (x - x_0) \qquad \forall x \in \Omega \\
&= u(x_1) - p \cdot (x_1 - x_0) + p \cdot (x - x_0) \\
&= u(x_1) + p \cdot (x - x_1) \qquad \forall x \in \Omega,
\end{aligned}$$

that is, $p \in \partial u(x_1)$ which is impossible because we removed S from $\partial u(K)$. Hence,

$$u(x) > u(x_0) + p \cdot (x - x_0) \qquad \forall x \in \overline{U}, x \neq x_0,$$

and since $\overline{U}$ is compact and $u_k \to u$ uniformly on $\overline{U}$, we have

$$u_k(x) \ge u_k(x_0) + p \cdot (x - x_0) + \epsilon$$

for all $k \geq k_0$ and all $x \in \overline{U}$ and some $\epsilon < 0$. Now let

$$\delta_k = \min_{x \in \overline{U}}\{u_k(x) - u_k(x_0) - p \cdot (x - x_0) - \epsilon\}.$$

This minimum is attained at some $x_k \in \overline{U}$. We claim that p is the slope of a supporting hyperplane to u_k at the point $(x_k, u(x_k))$. Indeed,

$$\delta_k = u_k(x_k) - u_k(x_0) - p \cdot (x_k - x_0) - \epsilon$$

and since $u_k(x) \geq u_k(x_0) + p \cdot (x - x_0) + \epsilon + \delta_k$ for all $x \in \overline{U}$, we have

$$u_k(x) \geq u_k(x_k) + p \cdot (x - x_k) \qquad \forall x \in \overline{U}. \tag{1.2.1}$$

Since u_k is convex in Ω and U is open, (1.2.1) holds for all $x \in \Omega$, that is $p \in \partial u_k(x_k)$ for all $k \geq k_0$. This implies that $p \in \liminf_{k \to \infty} \partial u_k(\overline{U})$. ∎

Lemma 1.2.3 *If u_k are convex functions in Ω such that $u_k \to u$ uniformly on compact subsets of Ω, then the associated Monge–Ampère measures Mu_k tend to Mu weakly, that is*

$$\int_{\Omega} f(x) \, dMu_k(x) \to \int_{\Omega} f(x) \, dMu(x),$$

for every f continuous with compact support in Ω.

1.3   Viscosity solutions

Definition 1.3.1 *Let $u \in C(\Omega)$ be a convex function and $f \in C(\Omega)$, $f \geq 0$. The function u is a viscosity subsolution (supersolution) of the equation $\det D^2 u = f$ in Ω if whenever convex $\phi \in C^2(\Omega)$ and $x_0 \in \Omega$ are such that $(u - \phi)(x) \leq (\geq) (u - \phi)(x_0)$ for all x in a neighborhood of x_0, then we must have*

$$\det D^2\phi(x_0) \geq (\leq) f(x_0).$$

Remark 1.3.2 We claim that if $u \in C(\Omega)$ is convex, $\phi \in C^2(\Omega)$ and $u - \phi$ has a local maximum at $x_0 \in \Omega$, then

$$D^2\phi(x_0) \geq 0.$$

In fact, since $\phi \in C^2(\Omega)$, we have

$$\phi(x) = \phi(x_0) + D\phi(x_0) \cdot (x - x_0) + \frac{1}{2}\langle D^2\phi(x_0)(x - x_0), x - x_0 \rangle + o(|x - x_0|^2).$$

Hence, for x close to x_0 we get

$$u(x) \leq \phi(x) + u(x_0) - \phi(x_0)$$
$$= u(x_0) + D\phi(x_0) \cdot (x - x_0)$$
$$+ \frac{1}{2} \langle D^2\phi(x_0)(x - x_0), x - x_0 \rangle + o(|x - x_0|^2).$$

Since u is convex, there exists p such that $u(x) \geq u(x_0) + p \cdot (x - x_0)$ for all $x \in \Omega$. Given $|w| = 1$ and $\rho > 0$ small, by letting $x - x_0 = \rho w$ we obtain

$$\rho \, p \cdot w \leq \rho \, D\phi(x_0) \cdot w + \frac{1}{2} \rho^2 \langle D^2\phi(x_0)w, w \rangle + o(\rho^2).$$

Dividing this expression by ρ, letting $\rho \to 0$ and noting that the resulting inequality holds for all $|w| = 1$ gives that $p = D\phi(x_0)$. Hence $\langle D^2\phi(x_0)w, w \rangle \geq 0$ and the claim is proved.

Remark 1.3.3 We show that we may restrict the class of test functions used in the definition of viscosity subsolution or supersolution to the class of strictly convex quadratic polynomials. We shall first prove that, if the statement giving a strictly convex quadratic polynomial ϕ and $x_0 \in \Omega$ such that $(u - \phi)(x) \leq (u - \phi)(x_0)$ for all x in a neighborhood of x_0 implies that

$$\det D^2\phi(x_0) \geq f(x_0),$$

then u is a viscosity subsolution of the equation $\det D^2u = f$ in Ω. To prove the remark, let $\phi \in C^2(\Omega)$ be convex such that $u - \phi$ has a local maximum at $x_0 \in \Omega$. We write

$$\phi(x) = \phi(x_0) + D\phi(x_0) \cdot (x - x_0)$$
$$+ \frac{1}{2} \langle D^2\phi(x_0)(x - x_0), x - x_0 \rangle + o(|x - x_0|^2)$$
$$= P(x) + o(|x - x_0|^2). \tag{1.3.1}$$

Let $\epsilon > 0$ and consider the quadratic polynomial $P_\epsilon(x) = P(x) + \epsilon|x - x_0|^2$. We have

$$D^2 P_\epsilon(x_0) = D^2 P(x_0) + 2\epsilon Id = D^2\phi(x_0) + 2\epsilon Id,$$

and so the polynomial P_ϵ is strictly convex. We have $\phi(x) - P_\epsilon(x) = o(|x - x_0|^2) - \epsilon|x - x_0|^2 \leq 0$ and so $\phi - P_\epsilon$ has a local maximum at x_0. Hence $u - P_\epsilon$ has a local maximum at x_0. Then $\det D^2 P_\epsilon(x_0) = \det (D^2\phi(x_0) + 2\epsilon Id) \geq f(x_0)$. By letting $\epsilon \to 0$, we obtain the desired inequality.

To prove the statement for supersolutions, let $\phi \in C^2(\Omega)$ be convex such that $u - \phi$ has a local minimum at x_0. If $D^2\phi(x_0)$ has some zero eigenvalue, then $\det D^2\phi(x_0) = 0 \leq f(x_0)$. If all eigenvalues of $D^2\phi(x_0)$ are positive and $P(x)$ is given by (1.3.1), then $P_\epsilon(x) = P(x) - \epsilon|x - x_0|^2$ is strictly convex for all $\epsilon > 0$ sufficiently small. Proceeding as before, we now get that $u - P_\epsilon$ has a local minimum at x_0 and consequently $\det D^2\phi(x_0) \leq f(x_0)$.

We now compare the two notions of solutions: generalized solutions and viscosity solutions.

Proposition 1.3.4 *If u is a generalized solution to $Mu = f$ with f continuous, then u is a viscosity solution.*

Proof. Let $\phi \in C^2(\Omega)$ be a strictly convex function such that $u - \phi$ has a local maximum at $x_0 \in \Omega$. We can assume that $u(x_0) = \phi(x_0)$, then $u(x) < \phi(x)$ for all $0 < |x - x_0| \leq \delta$. This can be achieved by adding $r|x - x_0|^2$ to ϕ and letting $r \to 0$ at the end.

Let $m = \min_{\delta/2 \leq |x-x_0| \leq \delta}\{\phi(x) - u(x)\}$. We have $m > 0$. Let $0 < \epsilon < m$ and consider the set

$$S_\epsilon = \{x \in B_\delta(x_0) : u(x) + \epsilon > \phi(x)\}.$$

If $\delta/2 \leq |x-x_0| \leq \delta$, then $\phi(x)-u(x) \geq m$ and so $x \notin S_\epsilon$. Hence $S_\epsilon \subset B_{\delta/2}(x_0)$. Let $z \in \partial S_\epsilon$. Then there exist $x_n \in S_\epsilon$ and $\bar{x}_n \notin S_\epsilon$ such that $x_n \to z$ and $\bar{x}_n \to z$. Hence $u + \epsilon = \phi$ on ∂S_ϵ. Since both functions are convex in S_ϵ by Lemma 1.4.1, we have that

$$\partial(u + \epsilon)(S_\epsilon) \subset \partial\phi(S_\epsilon).$$

Since u is a generalized solution, this implies that

$$\int_{S_\epsilon} f(x)\,dx \leq |\partial(u + \epsilon)(S_\epsilon)| \leq |\partial\phi(S_\epsilon)| = \int_{S_\epsilon} \det D^2\phi(x)\,dx.$$

By the continuity of f we obtain that $\det D^2\phi(x_0) \geq f(x_0)$.

A similar argument shows that u is a viscosity supersolution. ∎

We shall prove in Section 1.7 the converse of Proposition 1.3.4.

1.4 Maximum principles

In this section we prove two maximum principles and a comparison principle for the Monge–Ampère equation.

We begin with the following basic lemma.

Lemma 1.4.1 *Let $\Omega \subset \mathbb{R}^n$ be a bounded open set, and $u, v \in C(\overline{\Omega})$. If $u = v$ on $\partial\Omega$ and $v \geq u$ in Ω, then*

$$\partial v(\Omega) \subset \partial u(\Omega);$$

see Figure 1.2.

Proof. Let $p \in \partial v(\Omega)$. There exists $x_0 \in \Omega$ such that

$$v(x) \geq v(x_0) + p \cdot (x - x_0), \qquad \forall x \in \Omega.$$

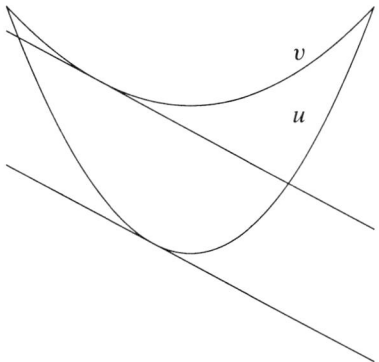

Figure 1.2. $\partial v(\Omega) \subset \partial u(\Omega)$

Let

$$a = \sup_{x \in \Omega}\{v(x_0) + p \cdot (x - x_0) - u(x)\}.$$

Since $v(x_0) \geq u(x_0)$, it follows that $a \geq 0$. We claim that $v(x_0) + p \cdot (x - x_0) - a$ is a supporting hyperplane to the function u at some point in Ω. Since Ω is bounded, there exists $x_1 \in \overline{\Omega}$ such that $a = v(x_0) + p \cdot (x_1 - x_0) - u(x_1)$ and so

$$u(x) \geq v(x_0) + p \cdot (x - x_0) - a = u(x_1) + p \cdot (x - x_1) \qquad \forall x \in \Omega.$$

We have

$$v(x_1) \geq v(x_0) + p \cdot (x_1 - x_0) = u(x_1) + a.$$

Hence, if $a > 0$, then $x_1 \notin \partial\Omega$ and so the claim holds in this case. If $a = 0$, then

$$u(x) \geq v(x_0) + p \cdot (x - x_0) \geq u(x_0) + p \cdot (x - x_0)$$

and consequently $u(x_0) + p \cdot (x - x_0)$ is a supporting hyperplane to u at x_0. ∎

1.4.1 Aleksandrov's maximum principle

The following estimate is fundamental in the study of the Monge–Ampère operator.

Theorem 1.4.2 (Aleksandrov's maximum principle) *If $\Omega \subset \mathbb{R}^n$ is a bounded, open and convex set with diameter Δ, and $u \in C(\overline{\Omega})$ is convex with $u = 0$ on $\partial\Omega$, then*

$$|u(x_0)|^n \leq C_n \Delta^{n-1} dist(x_0, \partial\Omega) |\partial u(\Omega)|,$$

for all $x_0 \in \Omega$, where C_n is a constant depending only on the dimension n.

Proof. Fix $x_0 \in \Omega$ and let v be the convex function whose graph is the upside-down cone with vertex $(x_0, u(x_0))$ and base Ω, with $v = 0$ on $\partial\Omega$. Since u is convex, $v \geq u$ in Ω. By Lemma 1.4.1

$$\partial v(\Omega) \subset \partial u(\Omega).$$

To prove the theorem, we shall estimate the measure of $\partial v(\Omega)$ from below. We first notice that the set $\partial v(\Omega)$ is convex. This is true because, if $p \in \partial v(\Omega)$, then there exists $x_1 \in \Omega$ such that $p = \partial v(x_1)$. If $x_1 \neq x_0$, since the graph of v is a cone, then $v(x_1) + p \cdot (x - x_1)$ is a supporting hyperplane at x_0, that is $p \in \partial v(x_0)$. So $\partial v(\Omega) = \partial v(x_0)$ and since $\partial v(x_0)$ is convex we are done.

Second, we notice that there exists $p_0 \in \partial v(\Omega)$ such that $|p_0| = \dfrac{-u(x_0)}{\text{dist}(x_0, \partial\Omega)}$. This follows because Ω is convex. Indeed, we take $x_1 \in \partial\Omega$ such that $|x_1 - x_0| = \text{dist}(x_0, \partial\Omega)$ and H is a supporting hyperplane to the set Ω at x_1. The hyperplane in $\mathbb{R}^{n+1}$ generated by H and the point $(x_0, u(x_0))$ is a supporting hyperplane to v that has the desired slope.

Now notice that the ball B with center at the origin and radius $\dfrac{-u(x_0)}{\Delta}$ is contained in $\partial v(\Omega)$, and $|p_0| \geq \dfrac{-u(x_0)}{\Delta}$. Hence the convex hull of B and p_0 is contained in $\partial v(\Omega)$ and it has measure

$$C_n \left(\frac{-u(x_0)}{\Delta} \right)^{n-1} |p_0|,$$

which proves the theorem.   ∎

Remark 1.4.3 The estimate in Theorem 1.4.2 is meaningful only if $|\partial u(\Omega)| = Mu(\Omega) < \infty$. If for example,

$$u(x) = \begin{cases} -\sqrt{x}, & \text{if } 0 \leq x \leq 1/2, \\ -\sqrt{1-x}, & \text{if } 1/2 \leq x \leq 1, \end{cases}$$

then $\partial u((0, 1)) = (-\infty, \infty)$. We also notice that an estimate of u in terms of the distance to the boundary of $(0, 1)$ is not valid.

1.4.2   Aleksandrov–Bakelman–Pucci's maximum principle

Consider $u \in C(\Omega)$ with Ω convex and the classes of functions

$$\mathcal{F}(u) = \{v : v(x) \leq u(x) \quad \forall x \in \Omega, \quad v \text{ convex in } \Omega\},$$

$$\mathcal{G}(u) = \{w : w(x) \geq u(x) \quad \forall x \in \Omega, \quad w \text{ concave in } \Omega\}.$$

Let

$$u_*(x) = \sup_{v \in \mathcal{F}(u)} v(x), \qquad u^*(x) = \inf_{w \in \mathcal{G}(u)} w(x). \tag{1.4.1}$$

We have that u_* is convex and u^* is concave in Ω. We call these functions *the convex and concave envelopes of u in Ω* respectively, and we have the inequalities

$$u_*(x) \le u(x) \le u^*(x), \qquad \forall x \in \Omega.$$

We also have that

$$\mathcal{F}(-u) = -\mathcal{G}(u)$$

and hence

$$-(u^*)(x) = (-u)_*(x). \tag{1.4.2}$$

Consider *the sets of contact points*

$$\mathcal{C}_*(u) = \{x \in \Omega : u_*(x) = u(x)\}, \qquad \mathcal{C}^*(u) = \{x \in \Omega : u^*(x) = u(x)\}.$$

Then

$$\mathcal{C}_*(u) = \mathcal{C}^*(-u). \tag{1.4.3}$$

Since u_* is convex, it follows that u_* has a supporting hyperplane at x_0, for $x_0 \in \mathcal{C}_*(u)$. Since in addition $u_*(x_0) = u(x_0)$, this hyperplane is also a supporting hyperplane to u at the same point. That is

$$\partial(u_*)(x_0) \subset \partial u(x_0), \qquad \text{for } x_0 \in \mathcal{C}_*(u),$$

and hence

$$\partial(u_*)(\mathcal{C}_*(u)) \subset \partial u(\mathcal{C}_*(u)).$$

If $x_0 \notin \mathcal{C}_*(u)$, then $\partial u(x_0) = \emptyset$. Also, if A, B are sets, then $\partial u(A \cup B) = \partial u(A) \cup \partial u(B)$. Hence

$$\partial u(\Omega) = \partial u(\mathcal{C}_*(u) \cup (\Omega \setminus \mathcal{C}_*(u))) = \partial u(\mathcal{C}_*(u)) \cup \partial u(\Omega \setminus \mathcal{C}_*(u)) = \partial u(\mathcal{C}_*(u)).$$

By the definition of u_*,

$$\partial u(\mathcal{C}_*(u)) \subset \partial(u_*)(\mathcal{C}_*(u)).$$

Thus

$$\partial u(\Omega) = \partial u(\mathcal{C}_*(u)) = \partial(u_*)(\mathcal{C}_*(u)). \tag{1.4.4}$$

Let

$$\Phi_u(x_0) = \{p : u(x) \le u(x_0) + p \cdot (x - x_0), \quad \forall x \in \Omega\}.$$

Notice that $\Phi_{-u}(x_0) = -\partial u(x_0)$.

Lemma 1.4.4 *Let $u \in C(\overline{\Omega})$ such that $u(x) \leq 0$ on $\partial\Omega$, and $x_0 \in \Omega$ with $u(x_0) >$
0. Then*

$$\Omega(x_0, u(x_0)) \subset \Phi_{u^*}(C^*(u)),$$

where $\Omega(x, t) = \{y : y \cdot (\xi - x) + t > 0, \forall \xi \in \overline{\Omega}\}$.

Proof. Let $y \in \Omega(x_0, u(x_0))$; then

$$y \cdot (\xi - x_0) + u(x_0) > 0, \quad \forall \xi \in \overline{\Omega}.$$

Let

$$\lambda_0 = \inf\{\lambda : \lambda + y \cdot (\xi - x_0) \geq u(\xi), \forall \xi \in \overline{\Omega}\}.$$

By continuity we have

$$\lambda_0 + y \cdot (\xi - x_0) \geq u(\xi), \qquad \forall \xi \in \overline{\Omega}. \tag{1.4.5}$$

Consider the minimum

$$\min_{\xi \in \overline{\Omega}} \{\lambda_0 + y \cdot (\xi - x_0) - u(\xi)\}.$$

This minimum is attained at some point $\overline{\xi} \in \overline{\Omega}$, and we have

$$\lambda_0 + y \cdot (\overline{\xi} - x_0) - u(\overline{\xi}) = 0.$$

Because on the contrary

$$\lambda_0 + y \cdot (\xi - x_0) - u(\xi) \geq \epsilon > 0, \qquad \forall \xi \in \overline{\Omega},$$

and λ_0 would not be the infimum.

We claim that $\overline{\xi} \in \Omega$. Since $u|_{\partial\Omega} \leq 0$, the claim will be proved if we show $u(\overline{\xi}) > 0$. By taking $\xi = x_0$ in (1.4.5) we get $u(x_0) \leq \lambda_0$, and consequently

$$y \cdot (\xi - x_0) + \lambda_0 > 0, \qquad \forall \xi \in \overline{\Omega};$$

in particular, for $\xi = \overline{\xi}$,

$$u(\overline{\xi}) = y \cdot (\overline{\xi} - x_0) + \lambda_0 > 0.$$

Therefore we proved that if $y \in \Omega(x_0, u(x_0))$, then there exists a point $\overline{\xi} \in \Omega$ such that

$$u(\overline{\xi}) = y \cdot (\overline{\xi} - x_0) + \lambda_0,$$

and

$$u(\xi) \leq y \cdot (\xi - x_0) + \lambda_0, \qquad \forall \xi \in \overline{\Omega}.$$

This means that $\lambda_0 + y \cdot (\xi - x_0)$ is a supporting hyperplane of u at $\overline{\xi}$. Since u^* is minimal, we have $u(\xi) \leq u^*(\xi) \leq y \cdot (\xi - x_0) + \lambda_0, \forall \xi \in \Omega$. In particular,

$u(\overline{\xi}) = u^*(\overline{\xi})$ and $y \cdot (\xi - x_0) + \lambda_0$ is a supporting hyperplane for u^* at $\overline{\xi}$, i.e., $y \in \Phi_{u^*}(\overline{\xi})$.   ∎

Then under the assumptions of Lemma 1.4.4 we get

$$\Omega(x_0, u(x_0)) \subset \Phi_{u^*}(C^*(u)) = -\partial(-(u^*))(C^*(u)) = -\partial((-u)_*)(C_*(-u)).$$

We also have the estimate

$$|\Omega(x_0, t)| \geq \frac{\omega_n t^n}{(\mathrm{diam}(\Omega))^n}. \tag{1.4.6}$$

To show (1.4.6), we first note that

$$\Omega(x_0, t) = t\Omega(x_0, 1),$$

which follows by writing $y \cdot (\xi - x_0) + t = t \left(\frac{y}{t} \cdot (\xi - x_0) + 1 \right)$. Also, if $x_0 \in \Omega$, then

$$B_{1/\mathrm{diam}(\Omega)}(0) \subset \Omega(x_0, 1). \tag{1.4.7}$$

In fact, let $\xi \in \overline{\Omega}$ and $y \in B_{1/\mathrm{diam}(\Omega)}(0)$ and write

$$\begin{aligned}
y \cdot (\xi - x_0) + 1 &= |y||\xi - x_0| \cos\phi + 1 \\
&= |y|\mathrm{diam}(\Omega) \frac{|\xi - x_0|}{\mathrm{diam}(\Omega)} \cos\phi + 1 \\
&\geq -|y|\mathrm{diam}(\Omega) \frac{|\xi - x_0|}{\mathrm{diam}(\Omega)} + 1 > 0,
\end{aligned}$$

and (1.4.7) follows.

Hence, if u satisfies the hypotheses of Lemma 1.4.4, then we get

$$\frac{\omega_n u(x_0)^n}{(\mathrm{diam}(\Omega))^n} \leq |-\partial((-u)_*)(C_*(-u))|.$$

We then obtain the following maximum principle:

Theorem 1.4.5 (Aleksandrov–Bakelman–Pucci's maximum principle) *If* $u \in C(\overline{\Omega})$ *and* $u \leq 0$ *on* $\partial\Omega$, *then*

$$\max_\Omega u(x) \leq \omega_n^{-1/n} \mathrm{diam}(\Omega) |\partial((-u)_*)(C_*(-u))|^{1/n}.$$

If in addition $u \in C^2(\Omega)$ *(without any assumptions on the sign of* u *on* $\partial\Omega$*), then*

$$\max_\Omega u(x) \leq \max_{\partial\Omega} u(x) + \omega_n^{-1/n} \mathrm{diam}(\Omega) \left(\int_{C_*(-u)} |\det D^2 u(x)| \, dx \right)^{1/n}.$$

Proof. It only remains to prove the last inequality. Subtracting from u the maximum on the boundary, we may assume that $u \leq 0$ on $\partial\Omega$. From (1.4.2), (1.4.3), and (1.4.4) we get $\partial((-u)_*)\,(C_*(-u)) = -\partial(-u)\,(C_*(-u))$. If $u \in C^2$ and $z \in C_*(-u)$, then $D^2(-u)(z) \geq 0$. Thus, by the formula for change of variables we obtain

$$|\partial(-u)\,(C_*(-u)))| \leq \int_{C_*(-u)} |\det D^2 u(x)|\, dx,$$

and the theorem is complete. ∎

1.4.3   Comparison principle

Theorem 1.4.6 *Let $u, v \in C(\overline{\Omega})$ be convex functions such that*

$$|\partial u(E)| \leq |\partial v(E)|, \qquad \text{for every Borel set } E \subset \Omega.$$

Then

$$\min_{x \in \overline{\Omega}}\{u(x) - v(x)\} = \min_{x \in \partial\Omega}\{u(x) - v(x)\}.$$

Proof. The proof is by contradiction. Let $a = \min_{x \in \overline{\Omega}}\{u(x) - v(x)\}$, $b = \min_{x \in \partial\Omega}\{u(x) - v(x)\}$, and assume $a < b$. There exists $x_0 \in \Omega$ such that $a = u(x_0) - v(x_0)$. Pick $\delta > 0$ sufficiently small such that $\delta\,(\operatorname{diam}\Omega)^2 < \dfrac{b-a}{2}$, and let

$$w(x) = v(x) + \delta|x - x_0|^2 + \frac{b+a}{2}.$$

Consider the set $G = \{x \in \overline{\Omega} : u(x) < w(x)\}$. We have $x_0 \in G$. Also, $G \cap \partial\Omega = \emptyset$. In fact, if $x \in G \cap \partial\Omega$, then $u(x) - v(x) \geq b$ and so

$$\begin{aligned}
w(x) &\leq u(x) + \delta|x - x_0|^2 - \frac{b-a}{2} \\
&\leq u(x) + \delta(\operatorname{diam}\Omega)^2 - \frac{b-a}{2} \\
&< u(x).
\end{aligned}$$

Hence $w(x) < u(x)$ for $x \in \partial\Omega$. This implies that $\partial G = \{x \in \Omega : u(x) = w(x)\}$. By Lemma 1.4.1 we obtain $\partial w(G) \subset \partial u(G)$. Also $\partial w = \partial(v + \delta|x - x_0|^2)$, and we have the inequality

$$|\partial(v + \delta|x - x_0|^2)(G)| \geq |\partial v(G)| + |\partial(\delta|x - x_0|^2)(G)|. \tag{1.4.8}$$

To prove (1.4.8), we observe that if A and B are symmetric and nonnegative definite matrices, then

$$\det(A + B) \geq \det A + \det B.$$

Hence, if $v \in C^2$, then (1.4.8) follows. In case v is not smooth, we can approximate v by a sequence $v_k \in C^2$ of convex functions converging uniformly on compact subsets of Ω. This can be achieved by taking a smooth function $\phi \geq 0$ with support in $B_1(0)$ and $\int \phi = 1$ and then letting $v_\epsilon = v \star \phi_\epsilon$. Hence (1.4.8) now follows from Lemma 1.2.2. Therefore

$$|\partial u(G)| \geq |\partial w(G)| \geq |\partial v(G)| + |\partial(\delta|x - x_0|^2)(G)| = |\partial v(G)| + (2\delta)^n |G|,$$

which contradicts the assumption of the theorem. ∎

Corollary 1.4.7 *If $u, v \in C(\overline{\Omega})$ are convex functions such that $|\partial u(E)| = |\partial v(E)|$ for every Borel set $E \subset \Omega$ and $u = v$ on $\partial\Omega$, then $u = v$ in Ω.*

1.5 The Dirichlet problem

Definition 1.5.1 *The open set $\Omega \subset \mathbb{R}^n$ is strictly convex if for all $x, y \in \overline{\Omega}$ the open segment joining x and y lies in Ω.*

Theorem 1.5.2 *Let $\Omega \subset \mathbb{R}^n$ be bounded and strictly convex, and $g : \partial\Omega \to \mathbb{R}$ a continuous function. There exists a unique convex function $u \in C(\overline{\Omega})$ generalized solution of the problem*

$$\det D^2 u = 0 \qquad in\ \Omega,$$
$$u = g \qquad on\ \partial\Omega.$$

Proof. Let $\mathcal{F} = \{a(x) : a \text{ is an affine function and } a \leq g \text{ on } \partial\Omega\}$. Since g is continuous, $\mathcal{F} \neq \emptyset$. Define

$$u(x) = \sup\{a(x) : a \in \mathcal{F}\}.$$

Since u is the supremum of convex functions, u is convex and $u(x) \leq g(x)$ for $x \in \partial\Omega$.

The first step is to show that $u = g$ on $\partial\Omega$. Let $\xi \in \partial\Omega$; we show that $u(\xi) \geq g(\xi)$. Given $\epsilon > 0$ there exists $\delta > 0$ such that $|g(x) - g(\xi)| < \epsilon$ for $|x - \xi| < \delta$, $x \in \partial\Omega$. Let $P(x) = 0$ be the equation of the supporting hyperplane to Ω at the point ξ, and assume that $\Omega \subset \{x : P(x) \geq 0\}$. Since Ω is strictly convex, there exists $\eta > 0$ such that $S = \{x \in \overline{\Omega} : P(x) \leq \eta\} \subset B_\delta(\xi)$. Let

$$M = \min\{g(x) : x \in \partial\Omega, P(x) \geq \eta\}$$

and consider

$$a(x) = g(\xi) - \epsilon - AP(x) \tag{1.5.1}$$

where A is a constant satisfying

$$A \geq \max\{\frac{g(\xi) - \epsilon - M}{\eta}, 0\}.$$

We have $a(\xi) = g(\xi) - \epsilon - AP(\xi) = g(\xi) - \epsilon$, and if $x \in \partial\Omega$ we claim that $a(x) \leq g(x)$. Indeed, if $x \in \partial\Omega \cap S$, then $g(\xi) - \epsilon \leq g(x) \leq g(\xi) + \epsilon$, so $g(x) \geq g(\xi) - \epsilon - AP(x) + AP(x) \geq g(\xi) - \epsilon - AP(x) = a(x)$. If $x \in \partial\Omega \cap S^c$, then $P(x) > \eta$ and by the definition of M and the choice of A we have

$$g(x) \geq M = a(x) + M - g(\xi) + \epsilon + AP(x)$$
$$\geq a(x) + M - g(\xi) + \epsilon + A\eta$$
$$\geq a(x).$$

Therefore $a \in \mathcal{F}$, and in particular $u(\xi) \geq a(\xi) = g(\xi) - \epsilon$ for every $\epsilon > 0$ and therefore $u(\xi) \geq g(\xi)$.

The second step is to show that u is continuous in $\overline{\Omega}$. Since u is convex in Ω, u is continuous in Ω. To prove the continuity on $\partial\Omega$, let $\xi \in \partial\Omega$, $\{x_n\} \subset \overline{\Omega}$ with $x_n \to \xi$. We show that $u(x_n) \to g(\xi)$. If a is the function constructed before, then $u(x) \geq a(x)$, in particular, $u(x_n) \geq a(x_n)$ and so $\liminf u(x_n) \geq \liminf a(x_n) = \liminf (g(\xi) - \epsilon - AP(x_n)) = g(\xi) - \epsilon$ for all $\epsilon > 0$. Hence $\liminf u(x_n) \geq g(\xi)$. We now prove that $\limsup u(x_n) \leq g(\xi)$. Since Ω is convex, there exists h harmonic in Ω such that $h \in C(\overline{\Omega})$ and $h|_{\partial\Omega} = g$. If a is any affine function so that $a \leq g$ on $\partial\Omega$, then a is harmonic and by the maximum principle $a \leq h$ in Ω. By taking supremum over a we obtain $u(x) \leq h(x)$ for $x \in \Omega$. In particular, $u(x_n) \leq h(x_n)$ and therefore $\limsup u(x_n) \leq \limsup h(x_n) = g(\xi)$ and we are done.

The third step is to prove that

$$\partial u(\Omega) \subset \{p \in \mathbb{R}^n : \text{there exist } x, y \in \Omega, x \neq y \text{ and } p \in \partial u(x) \cap \partial u(y)\},$$
$$(1.5.2)$$

and by Lemma 1.1.12, $|\partial u(\Omega)| = 0$.

If $p \in \partial u(\Omega)$, then there exists $x_0 \in \Omega$ such that $u(x) \geq u(x_0) + p \cdot (x - x_0) = a(x)$ for all $x \in \Omega$. Since $u = g$ on $\partial\Omega$, we have $g(x) \geq a(x)$ for all $x \in \partial\Omega$. There exists $\xi \in \partial\Omega$ such that $g(\xi) = a(\xi)$. Otherwise, there exists some $\epsilon > 0$ such that $g(x) \geq a(x) + \epsilon$ for all $x \in \partial\Omega$ and then $u(x) \geq a(x) + \epsilon$ for all $x \in \Omega$, and in particular $u(x_0) \geq a(x_0) + \epsilon = u(x_0) + \epsilon$, a contradiction. Since Ω is convex, the open segment I joining x_0 and ξ is contained in Ω. Now $u(x_0) = a(x_0)$ and $u(\xi) = a(\xi)$. If $z \in I$, then $z = tx_0 + (1-t)\xi$ and by convexity $u(z) \leq tu(x_0) + (1-t)u(\xi) = ta(x_0) + (1-t)a(\xi) = a(z)$. But $u(x) \geq a(x)$ for all $x \in \Omega$ so a is a supporting hyperplane to u at any point on the segment I, therefore $p \in \partial u(z)$ for all $z \in I$ and (1.5.2) is then proved.

Uniqueness follows from Corollary 1.4.7, but to illustrate we also include the following proof. Let $v \in C(\overline{\Omega})$, v convex and $v = g$ on $\partial\Omega$. Given $x_0 \in \Omega$, there exists a supporting hyperplane $a(x)$ at the point $(x_0, v(x_0))$, i.e., $v(x) \geq a(x)$ for all $x \in \overline{\Omega}$. Then $g(x) = v(x) \geq a(x)$ for $x \in \partial\Omega$, and so $a \in \mathcal{F}$ and $u(x) \geq a(x)$; in particular, $u(x_0) \geq a(x_0) = v(x_0)$. Therefore $u \geq v$ in $\overline{\Omega}$ and thus u is the largest convex function equal to g on $\partial\Omega$. To show that $u \leq v$, assume by contradiction that there exists $x_0 \in \Omega$ such that $u(x_0) > v(x_0)$. We shall show that this implies that $|\partial u(\Omega)| > 0$. Let $\epsilon = u(x_0) - v(x_0) > 0$ and let $a(x) = u(x_0) + p \cdot (x - x_0)$ be a supporting hyperplane to u at x_0, that is $u(x) \geq a(x)$ for all $x \in \Omega$. Consider the hyperplanes of the form $u(x_0) + q \cdot (x - x_0) - \dfrac{\epsilon}{2}$. We shall show that for q in a small ball around p this family of hyperplanes is below the graph of u. In fact, we have

$$u(x_0) + q \cdot (x - x_0) - \frac{\epsilon}{2}$$

$$= u(x_0) + p \cdot (x - x_0) + (q - p) \cdot (x - x_0) - \frac{\epsilon}{2}$$

$$\leq u(x_0) + p \cdot (x - x_0) + |q - p||x - x_0| - \frac{\epsilon}{2}$$

$$\leq u(x_0) + p \cdot (x - x_0) + \frac{\epsilon}{2} - \frac{\epsilon}{2}$$

$$\leq u(x),$$

for $|q - p| \leq \dfrac{\epsilon}{2M}$ where $M = \operatorname{diam}\Omega$. We now lower each of these hyperplanes until they become supporting hyperplanes to v at a certain point. The proof is similar to that of Lemma 1.4.1. One takes $a = \sup_{x \in \Omega}\{u(x_0) + q \cdot (x - x_0) - \frac{\epsilon}{2} - v(x)\}$, and we have $a > 0$ because at $x = x_0$ we have $u(x_0) - \frac{\epsilon}{2} - v(x_0) = \frac{\epsilon}{2} > 0$. Then there exists $x_1 \in \overline{\Omega}$ such that $a = u(x_0) + q \cdot (x_1 - x_0) - \frac{\epsilon}{2} - v(x_1)$, so $u(x_0) + q \cdot (x - x_0) - \frac{\epsilon}{2} - a \leq v(x)$, i.e., $u(x_0) + q \cdot (x - x_0) - \frac{\epsilon}{2}$ is a supporting hyperplane to v at x_1. It remains to show that $x_1 \in \Omega$. In fact, at x_1 we have $u(x_1) \geq u(x_0) + q \cdot (x_1 - x_0) - \frac{\epsilon}{2} = v(x_1) + a > v(x_1)$, and so $x_1 \notin \partial\Omega$. Consequently, $B_{\epsilon/2M}(p) \subset \partial v(\Omega)$ and the proof of the theorem is complete. ∎

Remark 1.5.3 The convex function $u(x, y) = \max(x^2 - 1, 0)$ is a generalized solution of $\det D^2 u = 0$ in $B_2(0, 0)$ that has continuous boundary data but is not regular because it has corners on the line $x = 1$.

1.6 The nonhomogeneous Dirichlet problem

In this section we solve the nonhomogeneous Dirichlet problem for the Monge–Ampère operator using the Perron method and Theorem 1.5.2. Let Ω be an open bounded and convex set, μ a Borel measure in Ω, and $g \in C(\partial\Omega)$. Set

$$\mathcal{F}(\mu, g) = \{v \in C\left(\bar{\Omega}\right) : v \text{ convex}, Mv \geq \mu \text{ in } \Omega, v = g \text{ on } \partial\Omega\}.$$

Suppose that $\mathcal{F}(\mu, g) \neq \emptyset$ and let $v \in \mathcal{F}(\mu, g)$. Assume that Ω is strictly convex. By Theorem 1.5.2, let $W \in C\left(\bar{\Omega}\right)$ be the unique convex solution of $MW = 0$ in Ω and $W = g$ on $\partial\Omega$. We have $0 = MW \leq \mu \leq Mv$ in Ω and by the comparison principle, Theorem 1.4.6, we have that $v \leq W$ in Ω. Therefore all functions in $\mathcal{F}(\mu, g)$ are uniformly bounded above and we can define

$$U(x) = \sup\{v(x) : v \in \mathcal{F}(\mu, g)\}. \tag{1.6.1}$$

The idea to solve the nonhomogeneous Dirichlet problem is first to construct U when the measure μ is a combination of delta masses, then to approximate a general measure μ by a sequence of measures of this form, and in this way construct the desired solution. With this in mind we need the following approximation lemma.

Lemma 1.6.1 *Let $\Omega \subset \mathbb{R}^n$ be a bounded open strictly convex domain, μ_j, μ be Borel measures in Ω, $u_j \in C(\bar{\Omega})$ convex, and $g \in C(\partial\Omega)$ such that*

1. $u_j = g$ on $\partial\Omega$,

2. $Mu_j = \mu_j$ in Ω,

3. $\mu_j \to \mu$ weakly in Ω, and

4. $\mu_j(\Omega) \leq A$ for all j.

Then $\{u_j\}$ contains a subsequence, also denoted by u_j, and there exists $u \in C(\bar{\Omega})$ convex in Ω such that u_j converges to u uniformly on compact subsets of Ω, and $Mu = \mu$, $u = g$ in $\partial\Omega$.

Proof. We have $u_j \in \mathcal{F}(\mu_j, g)$ and therefore u_j are uniformly bounded above. We prove that u_j are also uniformly bounded below in Ω. Let $\xi \in \partial\Omega$, $\epsilon > 0$, and $a(x) = g(\xi) - \epsilon - A P(x)$ be the affine function given by (1.5.1). Recall that $a(x) \leq g(x)$ for $x \in \partial\Omega$, $P(\xi) = 0$, $P(x) \geq 0$ for $x \in \Omega$, and $A \geq 0$. Set $v_j(x) = u_j(x) - a(x)$. If $x \in \partial\Omega$, then $v_j(x) = g(x) - a(x) \geq 0$, and the v_j are convex in Ω. If $v_j(x) \geq 0$ for all $x \in \Omega$, then u_j is bounded below in Ω. If at some point $v_j(x) < 0$, then by the Aleksandrov maximum principle, Theorem 1.4.2, applied to v_j on the set $G = \{x \in \Omega : v_j(x) \leq 0\}$, we obtain

$$(-v_j(x))^n \leq c_n \operatorname{dist}(x, \partial\Omega) \Delta^{n-1} Mv_j(\Omega) \leq c_n \operatorname{dist}(x, \partial\Omega) \Delta^{n-1} A,$$

with $\Delta = \operatorname{diam}(\Omega)$, and consequently $v_j(x) \geq -\left(c_n \operatorname{dist}(x, \partial\Omega) \Delta^{n-1} A\right)^{1/n}$, that is

$$u_j(x) \geq g(\xi) - \epsilon - A P(x) - C \left(\operatorname{dist}(x, \partial\Omega)\right)^{1/n}, \tag{1.6.2}$$

which proves that u_j are uniformly bounded below in Ω. On the other hand, $u_j(x) \leq w(x)$ with $\Delta w = 0$ in Ω and $w = g$ on $\partial \Omega$ by the maximum principle since u_j is weakly subharmonic. Now $\mathrm{dist}(x, \partial \Omega) \leq |x - \xi|$ and from (1.6.2) we obtain

$$w(x) \geq u_j(x) \geq g(\xi) - \epsilon - A\,P(x) - C\,|x - \xi|^{1/n}, \qquad (1.6.3)$$

and therefore $u_j(x) \to g(\xi)$ as $x \to \xi$.

Therefore by Lemma 1.1.6 and Lemma 3.2.1, we get that u_j are locally uniformly Lipschitz in Ω and by Arzèla–Ascoli there exists a subsequence, denoted also u_j, and a convex function u in Ω such that $u_j \to u$ uniformly on compact subsets of Ω. We also have from (1.6.3) that $u \in C(\bar{\Omega})$. The lemma then follows from Lemma 1.2.3. ∎

We now state and prove the main result in this section.

Theorem 1.6.2 *If $\Omega \subset \mathbb{R}^n$ is open bounded and strictly convex, μ is a Borel measure in Ω with $\mu(\Omega) < +\infty$, and $g \in C(\partial \Omega)$, then there exists a unique $u \in C(\bar{\Omega})$ that is a convex solution to the problem $Mu = \mu$ in Ω and $u = g$ on $\partial \Omega$.*

Proof. The uniqueness follows by the comparison principle, Theorem 1.4.6.

There exists a sequence of measures μ_j converging weakly to μ such that each μ_j is a finite combination of delta masses with positive coefficients and $\mu_j(\Omega) \leq A$ for all j. If we solve the Dirichlet problem for each μ_j with data g, then the theorem follows from Lemma 1.6.1. Therefore we assume from now on that

$$\mu = \sum_{i=1}^{N} a_i \, \delta_{x_i}, \qquad x_i \in \Omega, \qquad a_i > 0.$$

We claim that

(a) $\mathcal{F}(\mu, g) \neq \emptyset$.

(b) If $u, v \in \mathcal{F}(\mu, g)$, then $u \vee v \in \mathcal{F}(\mu, g)$.

(c) $U \in \mathcal{F}(\mu, g)$, with U defined by (1.6.1).

Step 1: proof of (a). By Example 1.1.16, $M(|x - x_i|) = \omega_n \, \delta_{x_i}$, with ω_n the volume of the unit ball in $\mathbb{R}^n$. Let $f(x) = \dfrac{1}{\omega_n^{1/n}} \sum_{i=1}^{N} a_i^{1/n}|x - x_i|$ and u be a solution to the Dirichlet problem $Mu = 0$ in Ω with $u = g - f$ on $\partial \Omega$. We claim that $v = u + f \in \mathcal{F}(\mu, g)$. Indeed, it is clear that $v \in C(\bar{\Omega})$, v is convex and $v = g$ on $\partial \Omega$. Let us calculate Mv. We have

$$Mv = M(u + f) \geq Mu + Mf \geq \frac{1}{\omega_n} \sum_{i=1}^{N} M\left(a_i^{1/n}|x - x_i|\right) = \sum_{i=1}^{N} a_i \, \delta_{x_i} = \mu.$$

Therefore $\mathcal{F}(\mu, g) \neq \emptyset$, and consequently U given by (1.6.1) is well defined.

Step 2: proof of (b). Let $\phi = u \vee v$, $\Omega_0 = \{x \in \Omega : u(x) = v(x)\}$, $\Omega_1 = \{x \in \Omega : u(x) > v(x)\}$, and $\Omega_2 = \{x \in \Omega : u(x) < v(x)\}$. If $E \subset \Omega_1$, then $M\phi(E) \geq Mu(E)$, and if $E \subset \Omega_2$, then $M\phi(E) \geq Mv(E)$. Also, if $E \subset \Omega_0$, then $\partial u(E) \subset \partial \phi(E)$ and $\partial v(E) \subset \partial \phi(E)$. Given $E \subset \Omega$ a Borel set, write $E = E_0 \cup E_1 \cup E_2$ with $E_i \subset \Omega_i$. We have

$$
\begin{aligned}
M\phi(E) &= M\phi(E_0) + M\phi(E_1) + M\phi(E_2) \\
&\geq Mu(E_0) + Mu(E_1) + Mv(E_2) \\
&\geq \mu(E_0) + \mu(E_1) + \mu(E_2) = \mu(E).
\end{aligned}
$$

Step 3: For each $y \in \Omega$ there exists a uniformly bounded sequence $v_m \in \mathcal{F}(\mu, g)$ converging uniformly on compact subsets of Ω to a function $w \in \mathcal{F}(\mu, g)$ so that $w(y) = U(y)$, where U is given by (1.6.1).

By Step 1, let $v_0 \in \mathcal{F}(\mu, g)$. If $v \in \mathcal{F}(\mu, g)$, then $v \leq W$ with W defined at the beginning of this section. Fix $y \in \Omega$, then by definition of U there exists a sequence $v_m \in \mathcal{F}(\mu, g)$ such that $v_m(y) \to U(y)$ as $m \to \infty$. Let $\bar{v}_m = v_0 \vee v_m$. By Step 2, $\bar{v}_m \in \mathcal{F}(\mu, g)$ and therefore $v_m(y) \leq \bar{v}_m(y) \leq U(y)$ and so $\bar{v}_m(y) \to U(y)$. Notice that $|\bar{v}_m(x)| \leq C_1$ for all $x \in \Omega$. Therefore we may assume that the original sequence v_m is bounded above and below in Ω. Since v_m is convex in Ω, it follows from Lemma 1.1.6 that given $K \subset \Omega$ compact, v_m is Lipschitz in K with constant

$$
C(K, m) = \sup\{|p| : p \in \partial v_m(K)\}.
$$

We claim that $C(K, m)$ is bounded uniformly in m. Let $p \in \partial v_m(x_0)$ with $x_0 \in K$. By Lemma 3.2.1, we get that $|p| \leq \dfrac{C_1}{\text{dist}(K, \Omega)}$ and the claim follows. Therefore v_m are equicontinuous on K and bounded in Ω. By Arzèla–Ascoli there exists a subsequence v_{m_j} converging uniformly on compact subsets of Ω to a function w, and so $w(y) = U(y)$. By Lemma 1.2.2 we have that $w \in \mathcal{F}(\mu, g)$ and hence $w \leq U$ in Ω.

Step 4: $MU \geq \mu$ in Ω. It is enough to prove that $MU(\{x_i\}) \geq a_i$ for $i = 1, \dots, N$. We may assume $i = 1$. By Step 3, there exists a sequence $v_m \in \mathcal{F}(\mu, g)$, uniformly bounded, such that $v_m \to w \in \mathcal{F}(\mu, g)$ uniformly on compacts of Ω as $m \to \infty$ with $w(x_1) = U(x_1)$. We have $Mw(\{x_1\}) \geq a_1$. If $p \in \partial w(x_1)$, then $w(x) \geq w(x_1) + p \cdot (x - x_1)$ in Ω and hence $U(x) \geq U(x_1) + p \cdot (x - x_1)$, that is $p \in \partial U(x_1)$. So $MU(\{x_1\}) = |\partial U(\{x_1\})| \geq |\partial w(\{x_1\})| \geq a_1$.

Step 5: $MU \leq \mu$ in Ω. We first prove that the measure MU is concentrated on the set $\{x_1, \dots, x_N\}$. Let $x_0 \in \Omega$ with $x_0 \neq x_i$, $i = 1, \dots, N$, and choose $r > 0$ so that $|x_i - x_0| > r$ for $i = 1, \dots, N$ and $B_r(x_0) \subset \Omega$. Solve $Mv = 0$ in $B_r(x_0)$ with $v = U$ on $\partial B_r(x_0)$, and define the "lifting of U"

$$
w(x) = \begin{cases} U(x) & x \in \Omega, |x - x_0| \geq r, \\ v(x) & |x - x_0| \leq r. \end{cases}
$$

We claim that $w \in \mathcal{F}(\mu, g)$. In fact, w is convex, because by Step 4, $MU \geq \mu \geq 0 = Mv$ in $B_r(x_0)$, and then by the comparison principle Theorem 1.4.6, $v \geq U$ in $B_r(x_0)$. It is clear that $w \in C\left(\overline{\Omega}\right)$. We verify that $Mw \geq \mu$ in Ω. Let $E \subset \Omega$ be a Borel set. We write

$$E = (E \cap B_r(x_0)) \cup \left(E \cap B_r(x_0)^c\right)$$

and so

$$Mw(E) = Mw\left(E \cap B_r(x_0)\right) + Mw\left(E \cap B_r(x_0)^c\right).$$

Now notice that if $F \subset B_r(x_0)$, then $\partial w(F) = \partial v(F)$, and if $F \subset B_r(x_0)^c$, then $\partial w(F) = \partial U(F)$. Therefore

$$Mw(E) = Mv\left(E \cap B_r(x_0)\right) + MU\left(E \cap B_r(x_0)^c\right) = 0 + MU\left(E \cap B_r(x_0)^c\right)$$
$$\geq \mu(E \cap B_r(x_0)^c) \geq \mu(E \cap \{x_1, \ldots, x_N\}) = \mu(E),$$

by (c) and the definition of μ. Therefore $w \leq U$, and since $w = v \geq U$ in $B_r(x_0)$, we get $v = U$ in $B_r(x_0)$, so $MU = Mv = 0$ in $B_r(x_0)$, where $B_r(x_0) \subset \Omega$ is any ball with $\overline{B_r(x_0)} \cap \{x_1, \ldots, x_N\} = \emptyset$. Hence if $E \subset \Omega$ is a Borel set with $E \cap \{x_1, \ldots, x_N\} = \emptyset$, then $MU(E) = 0$ by regularity of MU. Therefore MU is concentrated on the set $\{x_1, \ldots, x_N\}$, that is

$$MU = \sum_{i=1}^{N} \lambda_i \, a_i \, \delta_{x_i},$$

with $\lambda_i \geq 1$, $i = 1, \ldots, N$. We claim that $\lambda_i = 1$ for all $i = 1, \ldots, N$. Suppose by contradiction that $\lambda_i > 1$ for some i. Without loss of generality, we may assume that $MU = \lambda a \delta_0$, with $\lambda > 1$ and in the ball $B_r(0)$. We have $|\partial U(\{0\})| = \lambda a > 0$. Since $\partial U(\{0\})$ is convex, there exists a ball $B_\epsilon(p_0) \subset \partial U(\{0\})$. Then $U(x) \geq U(0) + p \cdot x$ for all $p \in B_\epsilon(p_0)$ and $x \in \Omega$. Let $V(x) = U(x) - p_0 \cdot x$. Then $V(x) \geq V(0) + (p - p_0) \cdot x$ for all $x \in \Omega$ and $p \in B_\epsilon(p_0)$. Given $x \in \Omega$ take $p - p_0 = \epsilon \, x/|x|$ and so

$$V(x) \geq V(0) + \epsilon |x|$$

for all $x \in \Omega$. Let α be a constant such that $V(0) - \alpha$ is negative and close to zero, and define $\bar{V}(x) = V(x) - \alpha$. We have $\bar{V}(0)$ is negative and small, and $\bar{V}(x) \geq \bar{V}(0) + \epsilon |x|$ for all $x \in \Omega$. If $r = -\dfrac{\bar{V}(0)}{\epsilon}$, then $\bar{V}(x) \geq \bar{V}(0) + \epsilon |x| \geq 0$ for all $|x| \geq r$. Let

$$w(x) = \begin{cases} \bar{V}(x) & \text{if } \bar{V}(x) \geq 0, \\ \lambda^{-1/n} \, \bar{V}(x) & \text{if } \bar{V}(x) < 0. \end{cases}$$

Notice that since $\lambda > 1$, we have $\lambda^{-1/n} \bar{V}(x) > \bar{V}(x)$ on the set $\{\bar{V}(x) < 0\}$. Consequently the function w is convex in Ω. Also, on the set $\{\bar{V}(x) < 0\}$, we have $Mw = M(\lambda^{-1/n}\,\bar{V}) = \frac{1}{\lambda}M\bar{V} = \frac{1}{\lambda}MU = a\,\delta_0$. On the other hand $w = \bar{V}$ on the set $\{\bar{V} \geq 0\}$, so $Mw = M\bar{V} = MU \geq \mu$ on the same set. Consequently, $Mw \geq \mu$ in Ω. This means that $w \in \mathcal{F}(\mu, \bar{g})$, where $\bar{g}$ are the boundary values of $\bar{V}(x) = U(x) - p_0 \cdot x - \alpha$. By definition of U,

$$\bar{V}(x) = U(x) - p_0 \cdot x - \alpha = \sup\{v(x) - p_0 \cdot x - \alpha : v \in \mathcal{F}(\mu, g)\}.$$

It is clear that $v'(x) = v(x) - p_0 \cdot x - \alpha \in \mathcal{F}(\mu, \bar{g})$ if and only if $v(x) \in \mathcal{F}(\mu, g)$. Therefore,

$$\bar{V}(x) = \sup\{v' : v' \in \mathcal{F}(\mu, \bar{g})\},$$

and since $w \in \mathcal{F}(\mu, \bar{g})$, we get that $w(x) \leq \bar{V}(x)$ for all $x \in \Omega$. In particular, $w(0) \leq \bar{V}(0)$ and so $\lambda^{-1/n}\,\bar{V}(0) \leq \bar{V}(0)$, and since $\bar{V}(0) < 0$ we obtain $\lambda^{-1/n} \geq 1$, a contradiction since $\lambda > 1$. This completes the proof of Step 5 and the theorem. ∎

1.7   Return to viscosity solutions

We prove here the following converse of Proposition 1.3.4.

Proposition 1.7.1 *Let $f \in C(\bar{\Omega})$ with $f > 0$ in $\bar{\Omega}$. If u is a viscosity solution to $\det D^2 u = f$ in Ω, then u is a generalized solution to $Mu = f$ in Ω.*

Proof. We have $0 < \lambda \leq f(x) \leq \Lambda$ in $\bar{\Omega}$. Given $x_0 \in \Omega$ and $0 < \eta < \lambda/2$, there exists $\epsilon > 0$ such that $f(x_0) - \eta < f(x) < f(x_0) + \eta$ for all $x \in B_\epsilon(x_0)$. Let $u_k \in C^\infty(\partial B_\epsilon(x_0))$ be a sequence such that $\max_{\partial B_\epsilon(x_0)} |u(x) - u_k(x)| \leq 1/k$, and v_k^+ and v_k^- the convex solutions to

$$\det D^2 v_k^{\pm} = f(x_0) \pm \eta, \qquad \text{in } B_\epsilon(x_0)$$
$$v_k^{\pm} = u_k, \qquad\qquad \text{on } \partial B_\epsilon(x_0).$$

We have that $v_k^{\pm} \in C^2(B_\epsilon(x_0)) \cap C(\bar{B}_\epsilon(x_0))$, see [GT83, section 17.7] or [CY77, Theorem 3, p. 59]; and

$$\det D^2 v_k^- < f(x) < \det D^2 v_k^+, \qquad \text{in } B_\epsilon(x_0) \text{ and}$$
$$u_k = v_k^{\pm}, \qquad\qquad \text{on } \partial B_\epsilon(x_0) \ .$$

By Lemma 1.7.2 below, we get

$$v_k^+(x) - \frac{1}{k} \leq u(x) \leq v_k^-(x) + \frac{1}{k} \quad \text{for } x \in \bar{B}_\epsilon(x_0). \tag{1.7.1}$$

By Theorem 1.6.2, let $v^{\pm}$ be the generalized solutions to

$$\det D^2 v^{\pm} = f(x_0) \pm \eta, \quad \text{in } B_\epsilon(x_0)$$
$$v^{\pm} = u, \qquad\qquad \text{on } \partial B_\epsilon(x_0) \ .$$

Applying the comparison principle Theorem 1.4.6, we get that $|v^{\pm}(x) - v_k^{\pm}(x)| \leq 1/k$ and consequently letting $k \to \infty$ in (1.7.1) yields

$$v^+(x) \leq u(x) \leq v^-(x) \qquad \text{for } x \in \bar{B}_\epsilon(x_0).$$

From Lemma 1.4.1 we obtain

$$\partial v^-(B_\epsilon(x_0)) \subset \partial u(B_\epsilon(x_0)) \subset \partial v^+(B_\epsilon(x_0)),$$

and consequently

$$|B_\epsilon(x_0)| \, (f(x_0) - \eta) \leq |\partial u(B_\epsilon(x_0))| = Mu(B_\epsilon(x_0)) \leq |B_\epsilon(x_0)| \, (f(x_0) + \eta). \tag{1.7.2}$$

Therefore if Q is a cube with diameter $\text{diam}(Q) < \epsilon$, then

$$C_1 |Q| \leq Mu(Q) \leq C_2 |Q|, \tag{1.7.3}$$

for some positive constants C_1, C_2. If $F \subset \Omega$ is a set of measure zero, then given $\delta > 0$ there exist a sequence of nonoverlapping cubes $Q_j \subset \Omega$ with $\text{diam}(Q_j) < \epsilon$, $F \subset \cup Q_j$, and $\sum |Q_j| < \delta$. Then applying (1.7.3) we obtain $Mu(F) < C_2 \delta$. That is, Mu is absolutely continuous with respect to Lebesgue measure and therefore there exists $h \in L^1_{\text{loc}}(\Omega)$ such that $Mu(E) = \int_E h(x) \, dx$. Dividing (1.7.2) by $|B_\epsilon(x_0)|$ and letting $\epsilon \to 0$ we get that $f(x_0) - \eta \leq h(x_0) \leq f(x_0) + \eta$ for almost all $x_0 \in \Omega$ and for all η sufficiently small. Hence Mu has density f. ∎

Lemma 1.7.2 *Suppose $f \in C(\Omega)$, $f \geq 0$, and $u \in C(\bar{\Omega})$ is a viscosity super-solution (subsolution) to $\det D^2 u = f$ in Ω. Suppose $v \in C^2(\Omega) \cap C(\bar{\Omega})$ is a classical convex solution to $\det D^2 v \geq (\leq)g$ in Ω with $g \in C(\Omega)$. If $f < (>)g$ in Ω, then*

$$\min_{\bar{\Omega}}(u - v) = \min_{\partial\Omega}(u - v) \qquad (\max_{\bar{\Omega}}(u - v) = \max_{\partial\Omega}(u - v)).$$

Proof. It is automatic from the definition (1.3.1). Suppose by contradiction that $\min_{\bar{\Omega}}(u - v) < \min_{\partial\Omega}(u - v)$. Then there exists $x_0 \in \Omega$ such that $(u - v)(x_0) = \min_{\bar{\Omega}}(u - v)$, and so $u - v$ has a local minimum at x_0. Since u is a viscosity supersolution to $\det D^2 u = f$ in Ω we get $g(x_0) \leq \det D^2 v(x_0) \leq f(x_0)$, a contradiction. ∎

1.8 Ellipsoids of minimum volume

An ellipsoid centered at a point x_0 is a set of the form

$$E(A, x_0) = \{x : \langle A(x - x_0), (x - x_0) \rangle \leq 1\}$$

where A is an $n \times n$ matrix that is symmetric and positive definite. The volume of $E(A, x_0)$ is

$$|E(A, x_0)| = \frac{\omega_n}{\sqrt{\det A}},$$

where ω_n is the volume of the unit ball in $\mathbb{R}^n$.

Lemma 1.8.1 *Let $S \subset \mathbb{R}^n$ be a bounded convex set.*

(a) *Assume that there exists $x_0 \in S$ such that $\overline{B_R(x_0)} \subset S$ and consider the class $\mathcal{F}_0$ of all ellipsoids with center at x_0 that contain the convex S. Then $\mathcal{F}_0$ has an ellipsoid of minimum volume.*

(b) *Assume that S has nonempty interior and consider the class $\mathcal{F}_1$ of all ellipsoids that contain the convex S. Then $\mathcal{F}_1$ has an ellipsoid of minimum volume.*

Proof. (a) Let $E(A, x_0)$ be an ellipsoid containing S, $A = (a_{ij})$. Then $\overline{B_R(x_0)} \subset E(A, x_0)$ and

$$|a_{ij}| \le \frac{1}{R^2}. \tag{1.8.1}$$

In fact, if ξ is a unit vector, then $x = x_0 + R\xi \in \overline{B_R(x_0)}$, and since $\overline{B_R(x_0)} \subset S$ we obtain

$$\langle A\xi, \xi \rangle \le \frac{1}{R^2}, \qquad \forall |\xi| = 1,$$

and (1.8.1) follows. Since $S \subset E$, we have $|E(A, x_0)| \ge |S| > 0$. Let $K = \{A \in \mathbb{R}^{n \times n} : S \subset E(A, x_0)\}$, and

$$\alpha = \inf_{A \in K} \frac{\omega_n}{\sqrt{\det A}}.$$

We have that $\alpha > 0$ and there exists a sequence $A_m = (a_{ij}^m) \in K$ such that $\dfrac{\omega_n}{\sqrt{\det A_m}} \to \alpha$. By (1.8.1) there exists a convergent subsequence $a_{ij}^{m_k} \to a_{ij}^0$ as $k \to \infty$. The matrix $A_0 = (a_{ij}^0)$ is symmetric and $A_0 \ge 0$. Since $\alpha > 0$, it follows that $\det A_0 > 0$ and then A_0 is positive definite. The desired ellipsoid is then $E(A_0, x_0)$.

(b) Let $E(A, x_1)$ be an ellipsoid containing S, $A = (a_{ij})$. Since S has nonempty interior there exists $B_R(x_2) \subset S$. Then $B_R(x_2) \subset E(A, x_1)$. Since $E(A, x_1)$ is an ellipsoid, $B_R(x_1) \subset E(A, x_1)$ and as before we get $|a_{ij}| \le \frac{1}{R^2}$. If $|x_1| \to \infty$, then $|E(A, x_1)| \to \infty$. Hence it is enough to consider $|x_1| \le M$ with M sufficiently large. Let $K' = \{(A, x_1) : S \subset E(A, x_1); |x_1| \le M\}$ and $\alpha = \inf_{K'} |E(A, x_1)| > 0$. By compactness we again obtain the desired ellipsoid. ∎

Theorem 1.8.2 *If $\Omega \subset \mathbb{R}^n$ is a bounded convex set with nonempty interior and E is the ellipsoid of minimum volume containing Ω centered at the center of mass of Ω, then*

$$\alpha_n E \subset \Omega \subset E,$$

where $\alpha_n = n^{-3/2}$ and αE denotes the α-dilation of E with respect to its center.

Proof. By using an affine transformation we may assume that E is the unit ball with center at the origin, and hence the center of mass of Ω is 0. Rotating the coordinates we may assume that $\mathrm{dist}(0, \partial\Omega) = \sigma = |x_0|$ with $x_0 = \sigma e_1 \in \partial\Omega$, $\sigma > 0$ and e_1 the coordinate unit vector in the direction $x_1 > 0$. Since Ω is convex, it follows that the plane $x_1 = \sigma$ is a supporting hyperplane to Ω at x_0. Moving the plane $x_1 = \sigma$ in a parallel fashion in the direction of x_1 negative, we obtain a plane Π that is a supporting hyperplane to Ω at a point $P \in \partial\Omega \cap \Pi$ and $\Pi = \{x_1 = -\mu\}$ for some $\mu > 0$. Consider the slice $S = \{x \in \Omega : x_1 = 0\}$, and let Γ be the cone with vertex P passing through S and contained in the slab $-\mu \leq x_1 \leq \sigma$; see Figure 1.3.

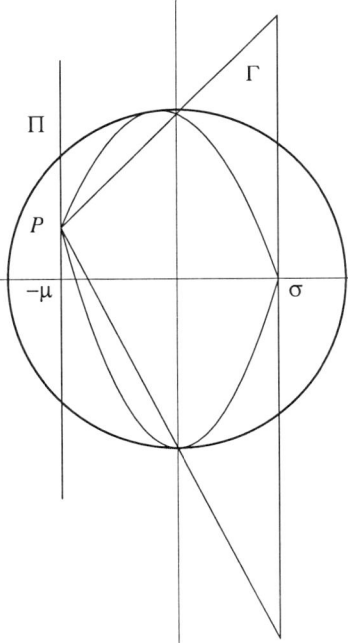

Figure 1.3. Theorem 1.8.2

The center of mass of Γ is

$$c(\Gamma) = \frac{1}{|\Gamma|} \int_\Gamma x \, dx.$$

Given $-\mu \le t \le \sigma$, let S_t be the slice of Γ passing through $(t, 0, \dots, 0)$ and perpendicular to the x_1-axis. The slice S_t is obtained by dilating S with respect to the point P, that is $S_t = \dfrac{t + \mu}{\mu} S$. Then by similarity, area$(S_t) = \left(\dfrac{t + \mu}{\mu}\right)^{n-1}$ area(S). Hence, if c_1 is the x_1-component of $c(\Gamma)$, then integrating on the slices we obtain

$$c_1 = \frac{1}{|\Gamma|} \text{area}(S) \int_{-\mu}^{\sigma} t \left(\frac{t + \mu}{\mu}\right)^{n-1} dt.$$

Notice that since Ω has center of mass 0, $\Gamma \cap \Omega$ has center of mass located to the right of S. Obviously, $\Gamma \cap \Omega \subset \Gamma$ and therefore the cone Γ also has center of mass to the right of S, that is $c_1 > 0$. Hence

$$\int_{-\mu}^{\sigma} t \left(\frac{t + \mu}{\mu}\right)^{n-1} dt > 0.$$

Changing variables and integrating this expression yields

$$\frac{\sigma}{\mu} \ge \frac{1}{n}. \tag{1.8.2}$$

Now consider

$$S_B = \{(x_1, x') : -\mu \le x_1 \le \sigma, |x'|^2 \le 1 - x_1^2\},$$

and the ellipsoid

$$E_0 = \{(x_1, x') : \frac{x_1^2}{a^2} + \frac{|x'|^2}{b^2} \le 1\},$$

where $\mu < a < 1 < b$. We claim that if $\mu < \dfrac{1}{\sqrt{n}}$, then there exist a and b such that $\Omega \subset S_B \subset E_0$ and $|E_0| < |B_1(0)|$. This contradicts the fact that E is the ellipsoid of minimum volume and therefore we must have $\mu \ge \dfrac{1}{\sqrt{n}}$. This inequality combined with (1.8.2) yields $\sigma \ge \dfrac{1}{n^{3/2}}$ and the theorem is proved.

To show the claim we need that $S_B \subset E_0$. Since $\mu \geq \sigma$, we have

$$\frac{x_1^2}{a^2} + \frac{|x'|^2}{b^2} \leq \frac{x_1^2}{a^2} + \frac{1 - x_1^2}{b^2}$$

$$= \frac{1}{b^2} + \left(\frac{1}{a^2} - \frac{1}{b^2} \right) x_1^2$$

$$\leq \frac{1}{b^2} + \left(\frac{1}{a^2} - \frac{1}{b^2} \right) \mu^2$$

$$= \frac{\mu^2}{a^2} + \frac{1 - \mu^2}{b^2}.$$

We have $S_B \subset E_0$ if $\dfrac{\mu^2}{a^2} + \dfrac{1 - \mu^2}{b^2} \leq 1$ which is equivalent to $b^2 \geq \dfrac{a^2(1 - \mu^2)}{a^2 - \mu^2}$.

Also, $|E_0| = ab^{n-1}|B_1(0)|$, and then $|E_0| < |B_1(0)|$ is equivalent to $ab^{n-1} < 1$. Then we want to choose a and b such that

$$\mu < a < 1 < b, \quad \text{and}$$

$$\frac{a^2(1 - \mu^2)}{a^2 - \mu^2} < b^2 < \left(\frac{1}{a} \right)^{2/(n-1)}. \tag{1.8.3}$$

We have $\dfrac{a^2(1 - \mu^2)}{a^2 - \mu^2} < \left(\dfrac{1}{a} \right)^{2/(n-1)}$ if and only if $a^2 - \mu^2 - a^{2n/(n-1)}(1 - \mu^2) > 0$. Consider the function $f(t) = t - \mu^2 - t^{n/(n-1)}(1 - \mu^2)$. We have $f(1) = 0$ and $f'(1) = 1 - \dfrac{n}{n-1}(1 - \mu^2)$. The assumption $\mu < \dfrac{1}{\sqrt{n}}$ is equivalent to $f'(1) < 0$. Hence $f(t) > 0$ for $t < 1$ and t near 1. By picking $t = a^2 < 1$ we obtain that $\dfrac{a^2(1 - \mu^2)}{a^2 - \mu^2} < \left(\dfrac{1}{a} \right)^{2/(n-1)}$ and hence we may choose $b^2 > 1$ satisfying (1.8.3). This proves the claim and hence the proof of the theorem is complete. ∎

1.9 Notes

The notions of normal mapping and generalized solutions to the Monge–Ampère equation as well as their properties, Lemma 1.1.12 and Theorem 1.1.13 are due to A. D. Aleksandrov; see [Pog64] and [Pog73]. The concept of viscosity solution is due to M. G. Crandall and P-L. Lions, see [CIL92]. The equivalence between the definitions of generalized and viscosity solutions proved in Propositions 1.3.4 and 1.7.1 is indicated in [Caf90a, pp. 137–139]. Theorem 1.4.2 is due to A. D. Aleksandrov, [Ale68]. The maximum principle Theorem 1.4.5 was discovered independently by A. D. Aleksandrov, I. Bakelman and C. Pucci; [Ale61], [Bak61], [Bak94], [Puc66]. Subsection 1.4.3, and Sections 1.5 and 1.6, are based on the paper [RT77]. The concept of ellipsoid of minimum volume is due to F. John, [Joh48], see also [dG75, p. 139], and Theorem 1.8.2 is a variant of a result there, see [Caf92, Lemma 2] and [Pog78, p. 91].

Chapter 2

Uniformly Elliptic Equations in Nondivergence Form

In this chapter we consider linear operators of the form

$$Lu = \sum_{i,j=1}^{n} a_{ij}(x) D_{ij} u(x)$$

where the coefficient matrix $A(x) = (a_{ij}(x))$ is symmetric and uniformly elliptic, that is

$$\lambda |\xi|^2 \leq \langle A(x)\xi, \xi \rangle \leq \Lambda |\xi|^2,$$

for all $\xi \in \mathbb{R}^n$ and $x \in \Omega \subset \mathbb{R}^n$. We assume that the coefficients a_{ij} are smooth functions, but the estimates we shall establish are independent of the regularity of the coefficients and depend only on the ellipticity constants λ, Λ and the dimension n.

2.1 Critical density estimates

Theorem 2.1.1 *There exist constants $M_0 > 1$ and $0 < \epsilon < 1$, depending only on the structure, such that for any $u \geq 0$ solution of $Lu \leq 0$ in the ball $B_{2R}(x_0)$ such that*

$$\inf_{B_R(x_0)} u \leq 1,$$

we have

$$|\{x \in B_{7R/4}(x_0) : u(x) < M_0\}| \geq \epsilon |B_{7R/4}(x_0)|.$$

Proof. Let $y \in B_{R/4}(x_0)$ and define the function

$$\phi_y(x) = \frac{R^2}{4} u(x) + \frac{|x-y|^2}{2}.$$

If $x \in B_R(x_0)$, then $|x - y| \leq \frac{5}{4}R$ and so

$$\phi_y(x) \leq \frac{R^2}{4}u(x) + \left(\frac{5}{4}R\right)^2 \frac{1}{2}, \qquad \text{for all } x \in B_R(x_0).$$

Therefore

$$\inf_{B_R(x_0)} \phi_y(x) \leq \frac{R^2}{4} + \left(\frac{5}{4}R\right)^2 \frac{1}{2} = \frac{33}{32}R^2.$$

On the other hand, if $x \in B_{2R}(x_0) \setminus B_{7R/4}(x_0)$, then $|x - y| \geq |x - x_0| - |x_0 - y| \geq \frac{6}{4}R$, and since $u \geq 0$,

$$\phi_y(x) \geq \frac{36}{32}R^2.$$

Therefore

$$\inf_{B_{2R}(x_0) \setminus B_{7R/4}(x_0)} \phi_y(x) \geq \frac{36}{32}R^2 > \frac{33}{32}R^2 \geq \inf_{B_R(x_0)} \phi_y(x),$$

and consequently

$$\inf_{B_{2R}(x_0)} \phi_y(x) = \inf_{B_{7R/4}(x_0)} \phi_y(x).$$

Hence, there exists $z \in B_{7R/4}(x_0)$ such that

$$\inf_{B_{2R}(x_0)} \phi_y(x) = \phi_y(z).$$

Let us consider the set

$$H = \{z \in B_{7R/4}(x_0) : \exists y \in B_{R/4}(x_0) \quad \text{such that} \quad \phi_y(z) = \inf_{B_{2R}(x_0)} \phi_y(x)\}.$$

That is, H is the set of all points $z \in B_{7R/4}(x_0)$ such that the minimum of $\phi_y(x)$ in $B_{2R}(x_0)$ is attained at z for some $y \in B_{R/4}(x_0)$. Now observe that

$$D\phi_y(z) = 0$$
$$D^2\phi_y(z) \geq 0.$$

Hence, $0 = D\phi_y(z) = \frac{R^2}{4}Du(z) + z - y$, which gives $y = \frac{R^2}{4}Du(z) + z$. Let us define the map

$$\Phi(z) = \frac{R^2}{4}Du(z) + z.$$

If $y \in B_{R/4}(x_0)$, then there exists $z \in H$ such that $\Phi(z) = y$. Then

$$|B_{R/4}(x_0)| \leq \int_{\Phi(H)} dx \leq \int_H |J_\Phi(x)|\, dx.$$

Now $J_\Phi(x) = \det\left(\dfrac{R^2}{4} D^2 u(x) + Id\right)$ and since $D^2\phi_y(x) \geq 0$ for $x \in H$, we get

$$D^2\phi_y(x) = \frac{R^2}{4} D^2 u(x) + Id \geq 0,$$

for $x \in H$. Therefore

$$|B_{R/4}(x_0)| \leq \int_H \det\left(\frac{R^2}{4} D^2 u(x) + Id\right) dx$$

$$= \int_H \frac{\det A(x)}{\det A(x)} \det\left(\frac{R^2}{4} D^2 u(x) + Id\right) dx$$

$$\leq n^{-n} \int_H \frac{1}{\det A(x)} \left(\text{trace}\left(A(x)\left(\frac{R^2}{4} D^2 u(x) + Id\right)\right)\right)^n dx$$

$$= n^{-n} \int_H \frac{1}{\det A(x)} \left(\left(\frac{R^2}{4} Lu(x) + \text{trace} A(x)\right)^+\right)^n dx.$$

Let us estimate the set H. Let $z \in H$; then there exists $y \in B_{R/4}(x_0)$ with

$$\phi_y(z) = \min_{x \in B_{2R}(x_0)} \phi_y(x).$$

Since $\inf_{B_R(x_0)} u \leq 1$, there exists $x_1 \in B_R(x_0)$ such that $u(x_1) \leq 1$. Then

$$\frac{R^2}{4} u(z) \leq \frac{R^2}{4} u(z) + \frac{|z - y|^2}{2} = \phi_y(z)$$

$$\leq \phi_y(x_1) = \frac{R^2}{4} u(x_1) + \frac{|x_1 - y|^2}{2}$$

$$\leq \frac{R^2}{4} + \left(\frac{5}{4} R\right)^2 \frac{1}{2} = \frac{33}{32} R^2.$$

Therefore, if $z \in H$, then $u(z) \leq 33/8$, that is

$$H \subset \{x \in B_{7R/4}(x_0) : u(x) \leq 33/8\}.$$

This yields the estimate

$$|B_{R/4}(x_0)|$$

$$\leq n^{-n} \int_{\{x \in B_{7R/4}(x_0): u(x) \leq 33/8\}} \frac{1}{\det A(x)} \left(\left(\frac{R^2}{4} Lu(x) + \text{trace} A(x)\right)^+\right)^n dx.$$

In particular, since $Lu \le 0$ we obtain

$$|B_{7R/4}(x_0)| \le 7^n n^{-n} \int_{\{x \in B_{7R/4}(x_0): u(x) \le 33/8\}} \frac{1}{\det A(x)} (\text{trace} A(x))^n \, dx,$$

and since the operator is strictly elliptic we obtain the estimate

$$|B_{7R/4}(x_0)| \le 7^n \left(\frac{\Lambda}{\lambda}\right)^n |\{x \in B_{7R/4}(x_0) : u(x) \le 33/8\}|$$

which completes the proof of the theorem.   ∎

Theorem 2.1.2 *There exists a constant $0 < \gamma \le 1$, depending only on the struc-ture, such that if $u > 0$ is any solution of $Lu \le 0$ in Ω with $u|_{B_R(x_0)} \ge 1$ and $B_{3R}(x_0) \subset \Omega$, then*

$$u|_{B_{2R}(x_0)} \ge \gamma.$$

Proof. Let $0 < \epsilon < 1$ and consider u^ϵ. We have

$$L(u^\epsilon) = \epsilon(\epsilon - 1)u^{\epsilon-2}\langle A Du, Du \rangle + \epsilon u^{\epsilon-1} Lu.$$

Hence, if $Lu(z) \le 0$, then by the ellipticity we get

$$L(-u^\epsilon)(z) \ge \lambda \epsilon(1 - \epsilon)u(z)^{\epsilon-2}|Du(z)|^2. \qquad (2.1.1)$$

Consider the function

$$h(x) = \frac{1}{8R^2}\left(4R^2 - |x - x_0|^2\right).$$

Then

$$0 \le h(x) \le \frac{1}{2}, \qquad \text{in } B_{2R}(x_0),$$

and since $Dh(x) = \frac{1}{4R^2}(x_0 - x)$,

$$\frac{1}{4R} \le |Dh(x)| \le \frac{1}{2R}, \qquad \text{for } R \le |x - x_0| \le 2R. \qquad (2.1.2)$$

Note also that $u(x) \ge h(x)$ in $B_R(x_0)$. Let

$$\delta = \min\{u(x)^\epsilon - h(x) : x \in B_{2R}(x_0)\},$$

where ϵ will be determined soon. By continuity, there exists $P \in \overline{B_{2R}(x_0)}$ such that

$$\delta = u(P)^\epsilon - h(P).$$

Case 1. Assume that $P \in B_{2R}(x_0) \setminus B_R(x_0)$. We then have

$$u(P)^{\epsilon} = \delta + h(P)$$
$$D(u^{\epsilon})(P) = Dh(P)$$
$$D^2(u^{\epsilon})(P) \geq D^2 h(P).$$

By (2.1.1) we have

$$L(-u^{\epsilon})(P) \geq \lambda \epsilon (1 - \epsilon) u(P)^{\epsilon - 2} |Du(P)|^2. \tag{2.1.3}$$

But

$$D(u^{\epsilon})(P) = \epsilon u(P)^{\epsilon - 1} Du(P) = Dh(P);$$

and

$$L(-u^{\epsilon})(P) = \text{trace}\left(A(P)\left(-D^2(u^{\epsilon})(P)\right)\right) \leq \text{trace}\left(A(P)\left(D^2(-h)(P)\right)\right).$$

From (2.1.3) we then get

$$\text{trace}\left(A(P)\left(D^2(-h)(P)\right)\right) \geq \lambda \left(\frac{1 - \epsilon}{\epsilon}\right) \frac{1}{u(P)^{\epsilon}} |Dh(P)|^2.$$

Since $D^2 h(x) = \dfrac{-1}{4R^2} Id$, it follows that

$$\text{trace}\left(A(P)\left(D^2(-h)(P)\right)\right) \leq \frac{n\Lambda}{4R^2}.$$

Combining these inequalities and (2.1.2) we obtain

$$u(P)^{\epsilon} \geq \left(\frac{\lambda}{\Lambda}\right) \frac{1}{4n} \left(\frac{1 - \epsilon}{\epsilon}\right),$$

which yields

$$\delta + h(P) \geq \left(\frac{\lambda}{\Lambda}\right) \frac{1}{4n} \left(\frac{1 - \epsilon}{\epsilon}\right).$$

Since $P \in B_{2R}(x_0) \setminus B_R(x_0)$, it follows that $h(P) \leq 3/8$. Therefore, by picking $\epsilon > 0$ sufficiently small we get

$$\delta \geq \left(\frac{\lambda}{\Lambda}\right) \frac{1}{4n} \left(\frac{1 - \epsilon}{\epsilon}\right) - \frac{3}{8} = C_{\epsilon} > 0.$$

Thus

$$u(x)^{\epsilon} \geq C_{\epsilon} + h(x) \geq C_{\epsilon},$$

which yields

$$u(x) \geq C_{\epsilon}^{1/\epsilon}, \qquad \text{for all } x \in B_{2R}(x_0).$$

Case 2. Assume that $P \in B_R(x_0)$. For this P we have $h(P) \leq \frac{1}{2}$, and so

$$\delta = u(P)^\epsilon - h(P) \geq 1 - h(P) \geq \frac{1}{2}.$$

Hence

$$u(x)^\epsilon \geq \frac{1}{2} + h(x) \geq \frac{1}{2},$$

and so

$$u(x) \geq \left(\frac{1}{2}\right)^{1/\epsilon}, \qquad \text{for all } x \in B_{2R}(x_0).$$

Case 3. Assume $P \in \partial B_{2R}(x_0)$. Since $h(P) = 0$ for $P \in \partial B_{2R}(x_0)$, $\delta = u(P)^\epsilon$. Hence

$$u(x)^\epsilon \geq u(P)^\epsilon + h(x) \geq h(x).$$

If $x \in B_{3R/2}(x_0)$, then $h(x) \geq 7/32$ and thus

$$u(x) \geq \left(\frac{7}{32}\right)^{1/\epsilon}, \qquad \text{for all } x \in B_{3R/2}(x_0).$$

Consequently, in any case we get

$$u(x) \geq \bar{C}_\epsilon > 0, \qquad \text{for all } x \in B_{3R/2}(x_0), \tag{2.1.4}$$

under the assumption $u \geq 1$ in $B_R(x_0)$.

Now, (2.1.4) implies that $\dfrac{u}{\bar{C}_\epsilon} \geq 1$ in $B_{3R/2}(x_0)$ and since $\dfrac{u}{\bar{C}_\epsilon}$ is also a positive supersolution of $Lu = 0$, by application of the previous argument we obtain

$$\frac{u}{\bar{C}_\epsilon} \geq \bar{C}_\epsilon, \qquad \text{for all } x \in B_{9R/4}(x_0),$$

and in particular $u \geq (\bar{C}_\epsilon)^2$ in $B_{2R}(x_0)$ and we are done. ∎

Theorem 2.1.3 *There exist constants $M_1 > 1$ and $0 < \epsilon < 1$, depending only on the structure, such that for any $u \geq 0$ that is a solution of $Lu \leq 0$ in the ball $B_{3R}(x_0)$ such that*

$$\inf_{B_{2R}(x_0)} u \leq 1,$$

we have

$$|\{x \in B_R(x_0) : u(x) < M_1\}| \geq \epsilon |B_R(x_0)|.$$

More generally, if $\inf_{B_{2^k R}(x_0)} u \leq 1$, then

$$|\{x \in B_R(x_0) : u(x) < M_0/\gamma^{k+1}\}| \geq \epsilon |B_R(x_0)|,$$

for $k = 1, 2, \ldots$, with M_0 and γ the constants in Theorems 2.1.1 and 2.1.2.

Proof. Let $M_1 = \dfrac{M_0}{\gamma^2}$ and assume that

$$|\{x \in B_R(x_0) : u(x) < M_1\}| < \epsilon|B_R(x_0)|.$$

This obviously implies that

$$|\{x \in B_R(x_0) : \gamma^2 u(x) < M_0\}| < \epsilon|B_R(x_0)|,$$

and since $\gamma^2 u$ is also a supersolution, by Theorem 2.1.1 it follows that

$$\inf_{B_{4R/7}(x_0)} \gamma^2 u \geq 1.$$

Hence, by Theorem 2.1.2,

$$\inf_{B_{8R/7}(x_0)} \gamma^2 u \geq \gamma,$$

i.e.,

$$\inf_{B_{8R/7}(x_0)} \gamma u \geq 1.$$

By applying again Theorem 2.1.2 to the supersolution γu, we obtain

$$\inf_{B_{16R/7}(x_0)} \gamma u \geq \gamma,$$

that is

$$\inf_{B_{16R/7}(x_0)} u \geq 1,$$

and since $B_{2R}(x_0) \subset B_{16R/7}(x_0)$, the proof of the theorem is complete. ∎

Remark 2.1.4 Theorem (2.1.3) implies a similar statement with cubes instead of balls and slightly different constants. This immediately follows noticing that if Q is a cube with center z and edge length ℓ, then $3Q \subset B(z, 3\sqrt{n}\ell/2)$. Hence if $\inf_{3Q} u \leq 1$, then applying Theorem (2.1.3) we get $|\{x \in Q : u(x) < M'\}| \geq \epsilon'|Q|$, where $M' > 1$ and $0 < \epsilon' < 1$.

2.2 Estimate of the distribution function of solutions

Let Q be a cube in R^n. We divide Q into 2^n congruent subcubes Q^k with disjoint interiors such that $|Q| = 2^{kn}|Q^k|$. Given a cube Q^k, its predecessor is denoted by $\tilde{Q}^k$, and Q^k is obtained from $\tilde{Q}^k$ by bisecting its sides, i.e., $|\tilde{Q}^k| = 2^n|Q^k|$.

Let A be a measurable set, and let Q be a cube in R^n. Suppose that $A \subset Q$ and

$$|A| \leq \delta|Q|,$$

for some $0 < \delta < 1$. The *Calderón–Zygmund decomposition of A at level δ* consists of a family $\{Q^k\}$ of dyadic subcubes of Q such that

$$A \subset \bigcup_{k=1}^{\infty} Q^k \qquad a.e.,$$

$$|A \cap Q^k| > \delta |Q^k|$$

and

$$|A \cap \tilde{Q}| \le \delta |\tilde{Q}|$$

for all $\tilde{Q}$ dyadic such that $\tilde{Q} \ne Q^k$ and $Q^k \subset \tilde{Q} \subset Q$.

Lemma 2.2.1 *Let $A \subset B \subset Q$ be measurable sets and Q be a cube. Assume that there exists $0 < \delta < 1$ such that*

(i) $|A| \le \delta |Q|$,

(ii) if Q^i is a dyadic cube of Q of the Calderón–Zygmund decomposition of A satisfying

$$|A \cap Q^i| > \delta |Q^i|,$$

we then have $\tilde{Q}^i \subset B$.

Then $|A| \le \delta |B|$.

Proof. We can select a subfamily of predecessors $\{\widetilde{Q}^{i_k}\}_{k=1}^{\infty}$ such that their interiors are disjoint and $\cup_{i=1}^{\infty} \tilde{Q}^i = \cup_{k=1}^{\infty} \tilde{Q}^{i_k}$. Hence $A \subset \cup_{k=1}^{\infty} \tilde{Q}^{i_k}$ a.e. and so

$$|A| \le \sum_{k=1}^{\infty} |A \cap \tilde{Q}^{i_k}| \le \delta \sum_{k=1}^{\infty} |\tilde{Q}^{i_k}| \le \delta |B|.$$

■

Theorem 2.2.2 *Let Q_0 be a cube such that $3Q_0 \subset \Omega$ and u a nonnegative solution of $Lu \le 0$ such that*

$$\inf_{3Q_0} u \le 1.$$

Then there exist positive constants $M > 1$ and $0 < \epsilon < 1$ depending only on the ellipticity constants such that

$$|\{x \in Q_0 : u(x) \ge M^k\}| \le (1 - \epsilon)^k |Q_0|, \qquad k = 1, \dots. \tag{2.2.1}$$

Proof. The proof is by induction. If $k = 1$, then the inequality follows from Remark 2.1.4. Assume that (2.2.1) holds for $k - 1$, and let

$$A = \{x \in Q_0 : u(x) \geq M^k\}, \qquad B = \{x \in Q_0 : u(x) \geq M^{k-1}\}.$$

Since $M > 1$, we have $A \subset B$ and by inductive hypothesis $|B| \leq (1 - \epsilon)^{k-1}|Q_0|$. We claim that

$$|A| \leq (1 - \epsilon)|B|.$$

To prove the claim we use Lemma 2.2.1. We need to show that (i) and (ii) of that lemma hold. Since $A \subset \{x \in Q_0 : u \geq M\}$, we have (i). To show (ii) let $\{Q^j\}$ be the Calderón–Zygmund decomposition of A at level $1 - \epsilon$. We want to show that the predecessor $\tilde{Q}^j$ of Q^j satisfies $\tilde{Q}^j \subset B$. Suppose by contradiction that this is not true, i.e., there exists a point $\bar{x} \in \tilde{Q}^j$ such that $\bar{x} \notin B$, that is $u(\bar{x}) < M^{k-1}$. Suppose Q^j has center x_j and edge length ℓ_j. Let $Q_1(0) = \{x : |x|_\infty \leq 1/2\}$, and set $Tx = x_j + \ell_j x$. We have $TQ_1(0) = Q^j$. Set

$$\bar{u}(x) = \frac{u(Tx)}{M^{k-1}}.$$

We have that $\bar{u}$ is a supersolution and $\tilde{Q}^j \subset TQ_3(0)$ where $Q_3(0) = \{x : |x|_\infty \leq 3/2\}$. Hence

$$\inf_{Q_3(0)} \bar{u} \leq 1,$$

and from Remark 2.1.4, $|\{x \in Q_1(0) : \bar{u}(x) \geq M\}| < (1 - \epsilon)$. Since

$$\{x \in Q_1(0) : \bar{u}(x) \geq M\} = T^{-1}\{x \in Q^j : u(x) \geq M^k\},$$

we obtain

$$|\{x \in Q^j : u(x) \geq M^k\}| < (1 - \epsilon)|Q^j|,$$

a contradiction. Then $\tilde{Q}^j \subset B$ and by Lemma 2.2.1 the proof is complete. ∎

Theorem 2.2.3 *Let $u \geq 0$ be a solution of $Lu \geq 0$ in Ω and assume that Q_1 is a cube with edge length one and Q_3 is a cube with edge length three concentric with Q_1, and $\inf_{Q_3} u \leq 1$. Let M and ϵ be the constants of Theorem 2.2.2.*

There exist positive constants k_0 and c depending only on the ellipticity constants with the following property:

If $k > k_0$ and $x_0 \in Q_1$ are such that

(1) $u(x_0) \geq M^k$,

(2) $d(x_0, Q_1^c) \geq c(1 - \epsilon)^{k/n}$,

then for $2\rho = c(1 - \epsilon)^{k/n}$ we have

$$\sup_{B_\rho(x_0)} u \geq u(x_0)(1 + \frac{1}{M}). \tag{2.2.2}$$

Proof.   By Theorem 2.2.2 we have

$$|A_1| = |\{x \in Q_1 : u(x) \geq M^{k-1}\}| \leq (1 - \epsilon)^{k-1}.$$

Suppose by contradiction that (2.2.2) is not true and consider

$$w(x) = \frac{u(x_0)(1 + \frac{1}{M}) - u(x)}{\frac{u(x_0)}{M}},$$

and the cube Q^* with center x_0 and edge length $\dfrac{\rho}{4\sqrt{n}}$. We have $\sup_{B_\rho(x_0)} u <$ $u(x_0)(1 + \frac{1}{M})$, and consequently $w(x) > 0$ in $B_\rho(x_0)$ and also $w(x_0) = 1$. Note also that $Q^* \subset B_\rho(x_0)$. We apply Theorem 2.2.2 to w in Q^* and we obtain

$$|A_2| = |\{x \in Q^* : w(x) \geq M\}| \leq (1 - \epsilon)|Q^*|.$$

We claim that $Q^* \subset A_1 \cup A_2$. In fact, if $x_1 \notin A_1$, then $u(x_1) < M^{k-1}$ and since $w(x) = M + 1 - \dfrac{Mu(x)}{u(x_0)}$, we have $w(x_1) > M + 1 - \dfrac{M^k}{u(x_0)}$. Then by (1) $w(x_1) \geq M$, i.e., $x_1 \in A_2$.

Consequently, $|Q^*| \leq |A_1| + |A_2| \leq (1 - \epsilon)^{k-1} + (1 - \epsilon)|Q^*|$, which implies

$$|Q^*| \leq \frac{(1 - \epsilon)^{k-1}}{\epsilon}.$$

Hence by the definition of ρ we get

$$\left(\frac{c}{8\sqrt{n}} (1 - \epsilon)^{k/n} \right)^n \leq \frac{(1 - \epsilon)^{k-1}}{\epsilon}.$$

This implies that

$$c \leq \frac{8\sqrt{n}}{((1 - \epsilon)\epsilon)^{1/n}}.$$

To obtain a contradiction we choose

$$c > \frac{8\sqrt{n}}{((1 - \epsilon)\epsilon)^{1/n}},$$

and k sufficiently large such that $c(1 - \epsilon)^{k/n}$ is sufficiently small.   ∎

2.3   Harnack's inequality

Theorem 2.3.1 *There exists a constant $C > 0$ depending only on the ellipticity constants and the dimension n, such that for any solution $u \geq 0$ of $Lu = 0$ in Ω and for any cube Q such that $3Q \subset \Omega$ we have*

$$\sup_Q u \leq C \inf_Q u.$$

Proof. By changing variables and rescaling, it is enough to prove the inequality for $Q = Q_1(0)$, the cube with edge length 1 centered at 0. We may also assume by dividing u by its minimum in Q that

$$\inf_{Q_1(0)} u = 1,$$

and we shall prove that

$$\sup_{Q_1(0)} u \leq C, \qquad (2.3.1)$$

for some constant C depending only on the structure. Given a cube Q_1 with edge length 1 such that $Q_1 \subset Q_2(0)$, ($Q_2(0)$ the cube with edge length 2 centered at 0) we have that $Q_1(0) \subset 2Q_1$ and so

$$\inf_{2Q_1} u \leq \inf_{Q_1(0)} = 1.$$

We shall prove that

$$u(x) \leq D_0 d(x, \partial Q_1)^{-\delta}, \qquad \text{for all } x \in Q_1, \qquad (2.3.2)$$

for some constants D_0 and δ depending only on the ellipticity constants and the dimension n. The inequality (2.3.2) blows up on the boundary of Q_1. By covering $Q_1(0)$ with a finite number of cubes $\frac{1}{2}Q_1$, with Q_1 as above, and applying (2.3.2) on each of these cubes we obtain (2.3.1).

Let M and ϵ be the constants of Theorem 2.2.3, and define $\delta > 0$ such that

$$\frac{1}{M} = (1 - \epsilon)^{\delta/n}.$$

Let

$$D = \sup_{x \in Q_1} u(x) d(x, \partial Q_1)^{\delta}.$$

Since u is continuous, there exists $x_0 \in Q_1$ such that

$$D = u(x_0) d(x_0, \partial Q_1)^{\delta}.$$

Pick an integer k such that

$$M^k < u(x_0) \leq M^{k+1}.$$

Hence

$$d(x_0, \partial Q_1) = \left(\frac{D}{u(x_0)}\right)^{1/\delta} \geq \left(\frac{D}{M^{k+1}}\right)^{1/\delta} = \left(\frac{D}{M}\right)^{1/\delta} (1 - \epsilon)^{k/n}. \qquad (2.3.3)$$

Let c and k_0 be the constants of Theorem 2.2.3. If $\dfrac{D}{M} < c^\delta$, then

$$u(x)\operatorname{dist}(x, \partial Q_1)^\delta \le D \le c^\delta M,$$

and (2.3.2) follows.

If $\dfrac{D}{M} \ge c^\delta$, then either $k > k_0$ or $k \le k_0$. If $k \le k_0$, then $u(x_0) \le M^{k_0+1}$, consequently $D \le M^{k_0+1}d(x_0, \partial Q_1)^\delta \le c_n M^{k_0+1}$ and (2.3.2) follows.

The worst case is then when

$$D \ge c^\delta M, \qquad k \ge k_0.$$

In this case the hypotheses of Theorem 2.2.3 are satisfied. We then have

$$\sup_{B_\rho(x_0)} u \ge u(x_0)(1 + \frac{1}{M}) = \frac{D}{d(x_0, \partial Q_1)^\delta}(1 + \frac{1}{M}). \qquad (2.3.4)$$

On the other hand, $B_\rho(x_0) \subset Q_1$ and consequently

$$\sup_{B_\rho(x_0)} u = \sup_{B_\rho(x_0)} \left(\frac{u(x)d(x, \partial Q_1)^\delta}{d(x, \partial Q_1)^\delta} \right)$$

$$\le \frac{1}{d(B_\rho(x_0), \partial Q_1)^\delta} \sup_{B_\rho(x_0)} \left(u(x)d(x, \partial Q_1)^\delta \right)$$

$$\le \frac{D}{(d(x_0, \partial Q_1) - \rho)^\delta}.$$

Now from the definition of ρ and (2.3.3) we get

$$d(x_0, \partial Q_1) \ge \left(\frac{D}{M} \right)^{1/\delta} \frac{2}{c}\rho,$$

which applied to the previous estimate gives

$$\sup_{B_\rho(x_0)} u \le \frac{D}{d(x_0, \partial Q_1)^\delta} \left(1 - \frac{c}{2}\left(\frac{M}{D} \right)^{1/\delta} \right)^{-\delta}. \qquad (2.3.5)$$

Comparing (2.3.4) and (2.3.5) yields

$$1 + \frac{1}{M} \le \left(1 - \frac{c}{2}\left(\frac{M}{D} \right)^{1/\delta} \right)^{-\delta}.$$

This implies that

$$D \le M \left(\frac{2}{c}\left(1 - \left(1 + \frac{1}{M} \right)^{-1/\delta} \right) \right)^{-\delta},$$

and the proof of (2.3.2) is complete.   ∎

2.4 Notes

The Harnack inequality for solutions of linear equations with nondivergence structure, Theorem 2.3.1, is due to N. V. Krylov and M. Safonov, [KS80]. The approach used in this chapter follows [Caf89], and [CG97]. The critical density Theorem 2.1.1 is proved without using the convex envelope and follows an idea of X. Cabré, [Cab97]. Theorem 2.1.2 is proved adapting to a simpler situation an argument used in [CG97, Theorem 2] for the linearized Monge–Ampère operator. This is the linear partial differential operator, in general nonuniformly elliptic, given by $Lu = \text{trace}\left(\Phi(x) D^2 u(x)\right)$ where $\Phi(x)$ is the matrix of cofactors of the Hessian $D^2\phi(x)$. In the paper [CG97], we developed a theory for L which is analogous to the De Giorgi–Nash–Moser theory for uniformly elliptic operators in divergence form and the theory for elliptic equations in nondivergence form exposed in this chapter. In [CG97], Euclidean balls are replaced by cross-sections of the convex function ϕ defined in Chapter 3, and Lebesgue measure by the Monge–Ampère measure $M\phi$.

Chapter 3

The Cross-sections of Monge–Ampère

3.1 Introduction

Let $\phi : \mathbb{R}^n \to \mathbb{R}$ be a convex function.

Definition 3.1.1 *Given* $t > 0$, *and* $\ell(x) = \phi(x_0) + p \cdot (x - x_0)$ *a supporting hyperplane to* ϕ *at* $(x_0, \phi(x_0))$, *a cross-section , or section, of* ϕ *at height* t *is the convex set*

$$S_\phi(x_0, p, t) = \{x \in \mathbb{R}^n : \phi(x) < \ell(x) + t\}.$$

If ϕ *is smooth, then* ℓ *is unique,* $\ell(x) = \phi(x_0) + D\phi(x_0) \cdot (x - x_0)$, *and we write*

$$S_\phi(x_0, t) = S_\phi(x_0, p, t).$$

We recall that $\partial\phi$ denotes the normal mapping of ϕ given by Definition 1.1.1.

To illustrate the notion of cross-section we give two simple examples. The first one is when the function ϕ is given by a paraboloid $\phi(x) = |x - y|^2$, $y \in \mathbb{R}^n$. In this case, it is easy to see that $S_\phi(x_0, t) = B_{\sqrt{t}}(x_0)$. The second example is given by the function $\phi(x) = h\,|x - y|$ whose graph is a cone, $y \in \mathbb{R}^n$ and $h \in \mathbb{R}$. Any supporting hyperplane to ϕ at the point $(y, 0)$ that is not parallel to any generator line of the cone gives rise to sections that are ellipses. On the other hand, if we take a supporting hyperplane at a point $(x, \phi(x))$ with $x \neq y$, then the corresponding sections are paraboloids of $n - 1$ dimensions and therefore the sections are unbounded sets.

Our purpose in this chapter is to analyze in detail several important geometric properties of the sections of the convex function ϕ when its associated Monge–Ampère measure $M\phi$, see Definition 1.1.1, satisfies a doubling condition. One of the main results in this chapter is a geometric characterization of Monge–Ampère

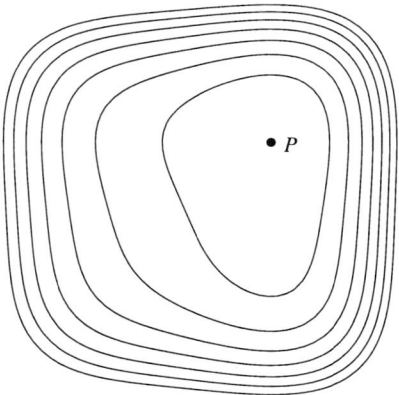

Figure 3.1. Cross-sections of $x^4 + y^4$ at $P = (2, 3/2)$

doubling measures, Theorem 3.3.5. This result is used to obtain estimates of the shape of the sections and invariance properties valid under appropriated normalizations. The properties established here will be used in subsequent chapters.

We shall assume throughout the chapter that the sections $S_\phi(x_0, p, t)$ are bounded sets. Let x_0^* be the center of mass of $S_\phi(x_0, p, t)$. If $\lambda > 0$, then $\lambda S_\phi(x_0, p, t)$ denotes the λ-dilation of $S_\phi(x_0, p, t)$ with respect to its center of mass, that is

$$\lambda S_\phi(x_0, p, t) = \{x_0^* + \lambda(x - x_0^*) : x \in S_\phi(x_0, p, t)\}.$$

We introduce the following two doubling conditions. We say that the Borel measure ν is *doubling with respect to the center of mass* on the sections of ϕ if there exist constants $C > 0$ and $0 < \alpha < 1$ such that for all sections $S_\phi(x, p, t)$,

$$\nu\left(S_\phi(x, p, t)\right) \leq C \nu\left(\alpha S_\phi(x, p, t)\right). \tag{3.1.1}$$

On the other hand, we say that ν is *doubling with respect to the parameter* on the sections of ϕ if there exists a constant $C' > 0$ such that for all sections $S_\phi(x, p, t)$,

$$\nu\left(S_\phi(x, p, t)\right) \leq C' \nu\left(S_\phi(x, p, t/2)\right). \tag{3.1.2}$$

Some comments about these doubling conditions are as follows. It will be shown that condition (3.1.1) implies (3.1.2), but the converse is in general false, see Corollary 3.3.2 and the subsequent remark. For examples of measures satisfying (3.1.1) see Remark 3.3.4. The assumption made that the sections $S_\phi(x, p, t)$ are bounded sets allows ϕ to have finite segments of lines in the graph. It is easy to see that if ϕ is strictly convex, then all the sections $S_\phi(x, p, t)$ are bounded sets, because otherwise the graph of ϕ may contain half-lines. As a consequence of Theorem 3.3.8, it follows that if the sections of ϕ are bounded sets and (3.1.1) holds, then ϕ is strictly convex, Remark 3.3.9.

We remark that in case the convex function ϕ is defined only on a convex open set $\Omega \subset \mathbb{R}^n$, then the results of this chapter hold true with straightforward modifications if we add to the hypothesis that the sections are bounded by the following condition: given $x \in \Omega$ there exists t_0 such that $\overline{S_\phi(x, p, t)} \subset \Omega$ for all $t \leq t_0$ and $p \in \partial\phi(x)$. The chapter is organized as follows. In Section 3.2, we present the basic facts needed in the proofs of the properties of the sections. Section 3.3 is subdivided into three parts. The characterization of Monge–Ampère doubling measures is contained in Section 3.3.1. Section 3.3.2 contains the proof of the engulfing property. A quantitative estimate of the size of the sections and some consequences are contained in Section 3.3.3.

3.2 Preliminary results

Let $T : \mathbb{R}^n \rightarrow \mathbb{R}^n$ be an invertible affine transformation, i.e., $Tx = Ax + b$ where A is an $n \times n$ invertible real matrix and $b \in \mathbb{R}^n$. If $\phi : \mathbb{R}^n \rightarrow \mathbb{R}$ and $\lambda > 0$, let $\psi_\lambda(y) = \dfrac{1}{\lambda}\phi(T^{-1}y)$. The function $\ell(x) = \phi(x_0) + p \cdot (x - x_0)$ is a supporting hyperplane of ϕ at the point $(x_0, \phi(x_0))$ if and only if $\bar{\ell}(y) = \psi(Tx_0) + \dfrac{1}{\lambda}(A^{-1})^t p \cdot (y - Tx_0)$ is a supporting hyperplane of the function ψ_λ at the point $(Tx_0, \psi_\lambda(Tx_0))$. Let μ and $\bar{\mu}$ be the Monge–Ampère measures associated with ϕ and ψ_λ respectively. That is

$$\mu(E) = |\partial\phi(E)|, \qquad \bar{\mu}(E) = |\partial\psi_\lambda(E)|.$$

Note that

$$\frac{1}{\lambda}\left(A^{-1}\right)^t (\partial\phi(E)) = \partial\psi_\lambda(TE),$$

and in case ϕ is regular,

$$D^2\psi_\lambda(x) = \frac{1}{\lambda}(A^{-1})^t (D^2\phi)(T^{-1}x) A^{-1}. \tag{3.2.1}$$

Consequently

$$\bar{\mu}(TE) = \frac{1}{\lambda^n}|\det A^{-1}|\mu(E). \tag{3.2.2}$$

In addition, the sections of ϕ and ψ_λ are related by the formula

$$T(S_\phi(x, p, t)) = S_{\psi_\lambda}(Tx, \frac{1}{\lambda}(A^{-1})^t p, \frac{t}{\lambda}). \tag{3.2.3}$$

Hence, noting that $T\left(\alpha S_\phi(x, p, t)\right) = \alpha T\left(S_\phi(x, p, t)\right)$, it follows that if μ satisfies either (3.1.1) or (3.1.2) on the sections of ϕ, then the measure $\bar{\mu}$ satisfies (3.1.1) or (3.1.2) on the sections of ψ_λ respectively and with the same constants.

We recall Theorem 1.8.2 and let E be the ellipsoid of minimum volume corresponding to Ω in that statement. There is an affine transformation T such that $T(E) = B(0, 1)$. Then

$$B(0, \alpha_n) \subset T(\Omega) \subset B(0, 1). \tag{3.2.4}$$

Here $B(x, t)$ denotes the Euclidean ball with center x and radius t. The set $T(\Omega)$ shall be called a *normalization* of Ω, and T shall be called an affine transformation that *normalizes* Ω. The center of mass of $T(\Omega)$ is 0 and by taking Lebesgue measure in (3.2.4), it follows that

$$\alpha_n^n \, \omega_n \frac{1}{|\Omega|} \leq |\det T| \leq \omega_n \frac{1}{|\Omega|}, \tag{3.2.5}$$

where ω_n denotes the volume of the unit ball in $\mathbb{R}^n$. We say that the convex set Ω is *normalized* when its center of mass is 0 and $B(0, \alpha_n) \subset \Omega \subset B(0, 1)$. If S is a section for the function ϕ then by (3.2.3) any normalization of S is also a section corresponding to the function ψ_λ.

We will also keep in mind the Aleksandrov maximum principle, Theorem 1.4.2.

The following lemmas give estimates of the size of the slopes of supporting hyperplanes to a convex function, [Caf92] and [CG97].

Lemma 3.2.1 *Let $\Omega \subset \mathbb{R}^n$ be a bounded convex open set and ϕ a convex function in Ω such that $\phi \leq 0$ on $\partial\Omega$. If $x \in \Omega$ and $\ell(y) = \phi(x) + p \cdot (y-x)$ is a supporting hyperplane to ϕ at the point $(x, \phi(x))$, then*

$$|p| \leq \frac{-\phi(x)}{dist(x, \partial\Omega)}. \tag{3.2.6}$$

More generally, if $\overline{\Omega'} \subset \Omega$, then

$$\partial\phi\left(\Omega'\right) \subset B\left(0, \frac{\max_{\Omega'}(-\phi)}{dist(\Omega', \partial\Omega)}\right). \tag{3.2.7}$$

Proof. Assume $p \neq 0$. We have $\phi(y) \geq \phi(x) + p \cdot (y - x)$, for all $y \in \Omega$. If $0 < r < \operatorname{dist}(x, \partial\Omega)$, then $y_0 = x + r\dfrac{p}{|p|} \in \Omega$ and $0 \geq \phi(y_0) \geq \phi(x) + r|p|$, which proves the lemma.   ∎

Lemma 3.2.2 *Let $\Omega \subset \mathbb{R}^n$ be a bounded convex open set, ϕ a convex function in Ω, $\phi = 0$ on $\partial\Omega$, and assume that the Monge–Ampère measure μ associated with ϕ satisfies $\mu(\Omega) < \infty$. Given $\lambda > 0$, let $\lambda\Omega$ denote the set obtained by λ-dilating*

Ω *with respect to its center of mass. There exists a positive constant* c_n *depending only on n such that*

$$B\left(0, \frac{|\min_\Omega \phi|}{2\operatorname{diam}(\Omega)}\right) \subset \partial\phi(\lambda\Omega),$$

for $\lambda_n < \lambda < 1$ *with*

$$\lambda_n = \max\left\{\frac{1}{2}, 1 - \frac{1}{c_n \mu(\Omega)}\left(\frac{|\min_\Omega \phi|}{\operatorname{diam}(\Omega)}\right)^n\right\}.$$

Proof. Let x_0 be the center of mass of Ω, and $\lambda\Omega = \{x_0 + \lambda(x - x_0) : x \in \Omega\}$. If $x \in \partial(\lambda\Omega)$, then $\operatorname{dist}(x, \partial\Omega) \le (1 - \lambda)\operatorname{diam}(\Omega)$. Thus, by the Aleksandrov estimate

$$|\phi(x)|^n \le c_n (\operatorname{diam}(\Omega))^n (1 - \lambda)\,\mu(\Omega),$$

for $x \in \partial(\lambda\Omega)$. We set $m_\lambda = \min_{\partial(\lambda\Omega)} \phi(x)$, $m = \min_\Omega \phi$, and we have $m = \phi(z_0)$ for some $z_0 \in \Omega$. Let us choose $1/2 \le \lambda \le 1$ and such that

$$c_n (\operatorname{diam}(\Omega))^n (1 - \lambda)\,\mu(\Omega) \le \left(\frac{-m}{2}\right)^n.$$

For this choice of λ we have $m_\lambda \ge \dfrac{m}{2}$. Now take the cylinder $\mathcal{C}$ in $\mathbb{R}^{n+1}$ perpendicular to $\mathbb{R}^n$ passing through the set $\partial(\lambda\Omega)$. Let $\mathcal{A} = \mathcal{C} \cap \{(x, \phi(x)) : x \in \Omega\}$ and v the convex function whose graph is the cone with vertex $(z_0, \phi(z_0))$ and passing through $\mathcal{A}$. We have that $v = \phi$ on $\partial(\lambda\Omega)$ and $v(x) \ge \phi(x)$ for $\lambda\Omega$. Therefore, by Lemma 1.4.1, we have that $\chi_v(\lambda\Omega) \subset \partial\phi(\lambda\Omega)$. Notice that $B\left(0, \dfrac{m_\lambda - m}{\operatorname{diam}(\lambda\Omega)}\right) \subset \chi_v(\lambda\Omega)$. Since $m_\lambda - m \ge \dfrac{-m}{2}$ and $\operatorname{diam}(\lambda\Omega) \le \operatorname{diam}(\Omega)$, it follows that $B\left(0, \dfrac{-m}{2\operatorname{diam}(\Omega)}\right) \subset \chi_v(\lambda\Omega)$ which proves the lemma. ∎

Combining Lemmas 3.2.1 and 3.2.2, we obtain the following proposition, see [Caf92] and Lemma 1.1 of [CG97].

Proposition 3.2.3 *Let* Ω *be a convex domain in* $\mathbb{R}^n$ *with center of mass equal* 0 *and* $B(0, \alpha_n) \subset \Omega \subset B(0, 1)$, *and* ϕ *a convex function in* Ω, *$\phi = 0$ on* $\partial\Omega$. *Let* μ *be the Monge–Ampère measure associated with* ϕ *and assume that there exist constants* $C > 0$ *and* $0 < \alpha < 1$ *such that* $\mu(\Omega) \le C\,\mu(\alpha\,\Omega)$. *Then*

$$C_1|\min_\Omega \phi|^n \le \mu(\Omega) \le C_2|\min_\Omega \phi|^n$$

where C_1, C_2 *are positive constants depending only on* C, α *and* n.

Proof. The first inequality follows from Lemma 3.2.2. By Lemma 3.2.1, it follows that

$$\partial\phi(\alpha\Omega) \subset B\left(0, \frac{\max_{\alpha\Omega} -\phi}{\operatorname{dist}(\alpha\Omega, \partial\Omega)}\right).$$

Hence

$$\mu(\alpha\Omega) \leq c_n \left(\frac{\max_{\alpha\Omega} -\phi}{\operatorname{dist}(\alpha\Omega, \partial\Omega)} \right)^n.$$

Notice that $\operatorname{dist}(\alpha\Omega, \partial\Omega) \geq \alpha_n (1 - \alpha)$. Indeed, there exist $x \in \alpha\Omega$ and $x' \in \partial\Omega$ such that $|x - x'| = \operatorname{dist}(\alpha\Omega, \partial\Omega)$. Let $z \in \partial\Omega$ be the point obtained by intersecting $\partial\Omega$ with the ray emanating from 0 and passing through x. If $|x - z| > |x - x'|$, then the angle between the segments $\overline{zx}$ and $\overline{zx'}$ is smaller than $\pi/2$. Let ℓ be the line joining z and x'. Since Ω is convex, the set $\ell \setminus \overline{zx'}$ does not intersect Ω, i.e., if $y \in \ell \setminus \overline{zx'}$, then $|y| \geq \alpha_n$. Let ℓ' be the ray emanating from 0 that is parallel to the segment $\overline{xx'}$ and lies on the plane containing z, x and x'. Let P be the intersection of ℓ' and ℓ. The claim now follows by similarity, considering the triangle with vertices x, z, x' and the triangle with vertices 0, z, P. We then obtain the second inequality. ∎

Corollary 3.2.4 *Let ϕ be convex in Ω with $0 < \lambda \leq M\phi \leq \Lambda$ in Ω. Suppose that $S_\phi(x_0, p, t) \subset \Omega$. Then $|S_\phi(x_0, p, t)| \approx t^{n/2}$.*

Proof. Let T be an affine transformation normalizing $S_\phi(x_0, p, t)$. Then by (3.2.3)

$$T(S_\phi(x_0, p, t)) = S_\psi(Tx_0, A^{-1}p, t) = \{y \in \mathbb{R}^n : \psi(y) < \bar{\ell}(y) + t\},$$

where $\psi(x) = \phi(T^{-1}x)$, and $\bar{\ell}(y) = \psi(Tx_0) + (A^{-1})^t p \cdot (y - Tx_0)$. Applying Proposition 3.2.3 to $h(y) = \psi(y) - \bar{\ell}(y) - t$ on the set $S_\psi(Tx_0, A^{-1}p, t)$, and using (3.2.2) and (3.2.4) we obtain the corollary. ∎

3.3 Properties of the sections

3.3.1 The Monge–Ampère measures satisfying (3.1.1)

The purpose of this section is to give a geometric characterization of the Monge–Ampère measures satisfying (3.1.1) and also to compare (3.1.1) and (3.1.2). We begin with

Lemma 3.3.1 *Let $0 < \lambda < 1$. Then*

$$\lambda S_\phi(x_0, p, t) \subset S_\phi\left(x_0, p, \left(1 - (1 - \lambda)\frac{\alpha_n}{2}\right)t\right).$$

Proof. Let T be an affine transformation that normalizes $S_\phi(x_0, p, t)$. We have $T\left(\lambda S_\phi(x_0, p, t)\right) = \lambda T\left(S_\phi(x_0, p, t)\right)$ and if $\psi(y) = \phi(T^{-1}y)$ and $q = (T^{-1})^t p$, then by (3.2.3) $T\left(\lambda S_\phi(x_0, p, t)\right) = \lambda S_\psi(Tx_0, q, t)$. We claim that

$$\lambda S_\psi(Tx_0, q, t) \subset S_\psi\left(Tx_0, q, \left(1 - (1 - \lambda)\frac{\alpha_n}{2}\right)t\right). \tag{3.3.1}$$

In fact, since $0 \in S_\psi(Tx_0, q, t)$ and by the convexity, there exists a point $\xi \in \partial S_\psi(Tx_0, q, t)$ and $0 < \theta \le 1$ such that $0 = \theta Tx_0 + (1 - \theta)\xi$. Hence, by (3.2.4)

$$\theta \ge \frac{|\xi|}{|Tx_0 - \xi|} \ge \frac{\alpha_n}{2}.$$

Let $\bar{\ell}(x) = \psi(Tx_0) + q \cdot (x - Tx_0)$ be the hyperplane defining $S_\psi(Tx_0, q, t)$. Since $\psi(y) - \bar{\ell}(y)$ is convex, it follows that

$$\psi(0) - \bar{\ell}(0) \le \theta \left(\psi(Tx_0) - \bar{\ell}(Tx_0)\right) + (1 - \theta) \left(\psi(\xi) - \bar{\ell}(\xi)\right) \le (1 - \frac{\alpha_n}{2})t.$$

Hence, if $y \in S_\psi(Tx_0, q, t)$, then

$$\psi(\lambda y) - \bar{\ell}(\lambda y) \le \lambda \left(\psi(y) - \bar{\ell}(y)\right) + (1 - \lambda) \left(\psi(0) - \bar{\ell}(0)\right)$$
$$\le \lambda t + (1 - \lambda)(1 - \frac{\alpha_n}{2})t$$
$$= \left(1 - (1 - \lambda)\frac{\alpha_n}{2}\right)t,$$

and the claim follows. The lemma follows by taking T^{-1} in (3.3.1). ∎

Corollary 3.3.2 ((3.1.1) implies (3.1.2)) *Let μ be a measure satisfying* (3.1.1). *Then μ satisfies* (3.1.2).

Proof. By Lemma 3.3.1

$$\mu(S_\phi(x, p, t)) \le C \mu(\alpha S_\phi(x, p, t)) \le C \mu(S_\phi(x, p, \theta t)), \qquad \theta = 1 - (1 - \alpha)\frac{\alpha_n}{2}.$$

By successive application of the last inequality

$$\mu(S_\phi(x, p, t)) \le C^k \mu(S_\phi(x, p, \theta^k t)),$$

and by taking k such that $\theta^k \le 1/2$, the corollary follows. ∎

Remark 3.3.3 The converse of Corollary 3.3.2 is false. The function $\phi(x) = e^x$, $x \in \mathbb{R}$, is strictly convex and the corresponding Monge-Ampère measure satisfies (3.1.2) but not (3.1.1). For such ϕ we observe that

(1) $S_\phi(x_0, t) = x_0 + S_\phi(0, te^{-x_0})$.

(2) $\mu\left(S_\phi(x_0, t)\right) = e^{x_0}|S_\phi(0, te^{-x_0})| = e^{x_0}\mu\left(S_\phi(0, te^{-x_0})\right)$.

Any interval of the form $(0, L)$, $L > 0$, is a section at some point. Given $0 < \alpha < 1$, $\alpha (0, L) = \frac{L}{2} + \alpha \left((0, L) - \frac{L}{2}\right) = \left(\frac{1 - \alpha}{2}L, \frac{1 + \alpha}{2}L\right)$. Thus

$$\frac{\mu\left((0, L)\right)}{\mu\left(\alpha(0, L)\right)} = \frac{e^L - 1}{e^{(1+\alpha)L/2} - e^{(1-\alpha)L/2}} \to \infty,$$

if $L \to \infty$ and hence (3.1.1) fails for any $0 < \alpha < 1$. Let us show that (3.1.2) holds for this measure. We have that $S_\phi(0, t) = (m_t, M_t)$ where $m_t < 0 < M_t$ and we have the following estimates:

(3) $\ln t \le M_t \le \ln 2t$ and $-2t \le m_t \le -t$ for $t \ge 5$.

(4) There exist positive constants ϵ and c such that $\epsilon \sqrt{t} \le M_t \le \sqrt{2t}$ and $-c\sqrt{t} \le m_t \le -\sqrt{2t}$ for $0 < t \le 5$.

We write

$$\frac{\mu\left(S_\phi(0, t)\right)}{\mu\left(S_\phi(0, t/2)\right)} = \frac{|S_\phi(0, t)|}{|S_\phi(0, t/2)|} = \frac{M_t - m_t}{M_{t/2} - m_{t/2}} = (*),$$

and from (3) and (4) it follows that $(*) \le C$ for all $t > 0$. This combined with (1) and (2) yields

$$\frac{\mu\left(S_\phi(x_0, t)\right)}{\mu\left(S_\phi(x_0, t/2)\right)} = \frac{e^{x_0}|S_\phi(0, t e^{-x_0})|}{e^{x_0}|S_\phi(0, t/2 e^{-x_0})|} \le C'.$$

Remark 3.3.4 If $p(x)$ is a polynomial in $\mathbb{R}^n$, then we shall show that the measure $|p(x)|\, dx$ satisfies (3.1.1) on the bounded sections of any convex function ϕ (actually on any convex set under the hypothesis of Theorem 1.8.2) and with a constant depending only on the degree of p. Since the polynomials of degree $\le d$ form a finite dimensional vector space, we have that

$$\int_{B_1(0)} p(x)^2\, dx \le C_d \left(\int_{B_1(0)} |p(x)|\, dx\right)^2,$$

for every polynomial p of degree $\le d$. Hence by Schwartz's inequality

$$\int_{B_1(0) \setminus B_{1-\epsilon}(0)} |p(x)|\, dx \le \omega_n (1 - (1-\epsilon)^n)^{1/2} C_d \int_{B_1(0)} |p(x)|\, dx$$

$$\le C\epsilon^{1/2} \int_{B_1(0)} |p(x)|\, dx,$$

with $C = C(n, d)$. Thus

$$\int_{B_1(0)} |p(x)|\, dx \le 2 \int_{B_{1-\epsilon}(0)} |p(x)|\, dx,$$

for ϵ sufficiently small. By changing variables

$$\int_{B_r(0)} |p(x)|\, dx \le 2 \int_{B_{(1-\epsilon)r}(0)} |p(x)|\, dx,$$

for all $r > 0$, and by iteration it follows that

$$\int_{B_1(0)} |p(x)|\, dx \leq C \int_{B_{\alpha_n/2}(0)} |p(x)|\, dx,$$

where α_n is the constant in Theorem 1.8.2. Now, let S be a section of ϕ and T an affine transformation that normalizes S, i.e., $B_{\alpha_n}(0) \subset S^* = T(S) \subset B_1(0)$. We then have

$$\int_S |p(x)|\, dx = \int_{S^*} |p(T^{-1}y)|\, |\det T^{-1}|\, dy \leq |\det T^{-1}| \int_{B_1(0)} |p(T^{-1}y)|\, dy$$

$$\leq C|\det T^{-1}| \int_{B_{\alpha_n/2}(0)} |p(T^{-1}y)|\, dy$$

$$\leq C|\det T^{-1}| \int_{\frac{1}{2}T(S)} |p(T^{-1}y)|\, dy = C \int_{\frac{1}{2}S} |p(x)|\, dx,$$

since $T(\frac{1}{2}S) = \frac{1}{2}T(S)$. This completes the remark.

The following theorem is one of the main results in the chapter and gives a geometric characterization of doubling Monge–Ampère measures.

Theorem 3.3.5 *Let μ be the Monge–Ampère measure associated with the convex function ϕ. The following statements are equivalent:*

(i) μ satisfies the doubling condition (3.1.1).

(ii) There exist $0 < \tau,\ \lambda < 1$ such that for all $x_0 \in \mathbb{R}^n$ and $t > 0$

$$S_\phi(x_0, p, \tau t) \subset \lambda S_\phi(x_0, p, t). \tag{3.3.2}$$

Proof. Let us assume (i). We shall show that there exists a dimensional constant $0 < \beta_n \leq 1$ such that (3.3.2) holds for all τ and λ such that $0 < \tau < 1$ and $1 - \beta_n(1 - \tau)^n \leq \lambda < 1$. Let T be an affine transformation that normalizes $S_\phi(x_0, p, t)$, x_0^* be the center of mass, and $\psi(y) = \phi(T^{-1}y)$. Let $0 < \lambda < 1$. By (3.2.3) we have $T\left(S_\phi(x_0, p, \lambda t)\right) = S_\psi(Tx_0, q, \lambda t)$, where $q = (T^{-1})^t p$. Since the center of mass of $S_\psi(Tx_0, q, \lambda t)$ is $Tx_0^* = 0$, we have

$$T\left(\lambda S_\phi(x_0, p, t)\right) = T\{x_0^* + \lambda(x - x_0^*) : x \in S_\phi(x_0, p, t)\}$$
$$= \{\lambda Tx : x \in S_\phi(x_0, p, t)\} = \lambda S_\psi(Tx_0, q, t).$$

Set $\psi^*(y) = \psi(y) - \psi(Tx_0) - q \cdot (y - Tx_0) - t$. Then $\partial \psi^* = \partial \psi - q$ and $\psi^*|_{\partial S_\psi(Tx_0, q, t)} = 0$. If $y \in S_\psi(Tx_0, q, t) \setminus \lambda\, S_\psi(Tx_0, q, t)$, then

$$\text{dist}(y, \partial S_\psi(Tx_0, q, t)) \leq 1 - \lambda,$$

and by Aleksandrov's estimate Theorem 1.4.2 and Proposition 3.2.3,

$$|\psi^*(y)|^n \leq c_n \text{dist}(y, \partial S_\psi(Tx_0, q, t))| \min_{S_\psi(Tx_0,q,t)} \psi^*(y)|^n \leq c_n(1-\lambda)t^n.$$

Hence $\psi^*(y) \geq -c_n^{-1/n}(1-\lambda)^{1/n}t$ which implies

$$\psi(y) - \psi(Tx_0) - q \cdot (y - Tx_0) \geq \left(1 - c_n^{-1/n}(1-\lambda)^{1/n}\right)t \geq \tau t,$$

that is, $y \notin S_\psi(Tx_0, q, \tau t)$. Therefore

$$S_\psi(Tx_0, q, \tau t) \subset \lambda S_\psi(Tx_0, q, t).$$

Hence, (3.3.2) follows by taking T^{-1}.

We now prove that (ii) implies (i). Let T be an affine transformation normalizing $S_\phi(x_0, p, t)$ and $\psi(y) = (\phi(T^{-1}y) - \phi(x_0) - q \cdot (y - Tx_0))/t$. Obviously, by (3.3.2) we have

$$S_\psi(Tx_0, 0, \tau) \subset \lambda S_\psi(Tx_0, 0, 1).$$

If $x \in S_\psi(Tx_0, 0, \tau)$ and q^* is the slope of a supporting hyperplane to ψ at $(x, \psi(x))$, then by Lemma 3.2.2,

$$|q^*| \leq \frac{1}{\text{dist}(x, \partial S_\psi(Tx_0, 0, 1))} \leq C_{n,\lambda},$$

and hence

$$\bar{\mu}\left(S_\psi(Tx_0, 0, \tau)\right) \leq C'_{n,\lambda}. \tag{3.3.3}$$

On the other hand, by applying Aleksandrov's estimate to $\psi(y) - \frac{\tau}{2}$ in $S_\psi(Tx_0, 0, \frac{\tau}{2})$ it follows that

$$\left(\frac{\tau}{2}\right)^n \leq C_n \bar{\mu}\left(S_\psi(Tx_0, 0, \tau/2)\right). \tag{3.3.4}$$

From (3.3.3) and (3.3.4) we obtain

$$\bar{\mu}\left(S_\psi(Tx_0, 0, \tau)\right) \leq C\bar{\mu}\left(S_\psi(Tx_0, 0, \tau/2)\right),$$

which implies

$$\mu\left(S_\phi(x_0, p, \tau t)\right) \leq C\mu\left(S_\phi(x_0, p, \tau t/2)\right),$$

where the constant C is independent of t. If we pick k such that $2^{-k} < \tau$, then by iteration we obtain

$$\mu\left(S_\phi(x_0, p, t)\right) \leq C\mu\left(S_\phi(x_0, p, t/2)\right)$$
$$\leq C^k\mu\left(S_\phi(x_0, p, 2^{-k}t)\right) \leq C'\mu\left(S_\phi(x_0, p, \tau t)\right).$$

Now (i) follows from (3.3.2). ∎

As a first consequence of our characterization we obtain

Corollary 3.3.6 *Let μ be the Monge-Ampère measure associated with the convex function ϕ and assume that μ satisfies (3.1.1). Let T be an affine transformation that normalizes the section $S_\phi(x, p, t)$, (in particular, by (3.2.3) $T\left(S_\phi(x, p, t)\right) = S_\psi(Tx, q, t)$ where $\psi(y) = \phi(T^{-1}y)$, and $q = (T^{-1})^t p$). Then*

(i) *There exists $c_0 > 0$ depending only on the constant in (3.1.1) and n such that*

$$\text{dist}\left(S_\psi(Tx, q, \tau t), \partial S_\psi(Tx, q, t)\right) \geq c_0(1 - \tau)^n, \qquad \text{for all } 0 < \tau < 1.$$

(ii) *There exists $C > 0$ depending only on the constant in (3.1.1) and n such that if $y \notin S_\phi(x, p, t)$ then*

$$B(T(y), C\epsilon^n) \cap T\left(S_\phi(x, p, (1 - \epsilon)t)\right) = \emptyset, \qquad \text{for all } 0 < \epsilon < 1. \tag{3.3.5}$$

Proof. (i). By Theorem 3.3.5, $S_\psi(Tx, q, \tau t) \subset \lambda S_\psi(Tx, q, t)$ with $\lambda = 1 - c_n(1 - \tau)^n$. Hence

$$\text{dist}\left(\lambda S_\psi(Tx, q, t), \partial S_\psi(Tx, q, t))\right) \geq \alpha_n(1 - \lambda) = c_n'(1 - \tau)^n,$$

and (i) follows.

(ii). By (i) we have

$$
\begin{aligned}
\text{dist}(T(S_\phi(x, p, (1 - \epsilon)t), T(y)) &= \text{dist}(S_\psi(Tx, q, (1 - \epsilon)t), T(y)) \\
&\geq \text{dist}(S_\psi(Tx, q, (1 - \epsilon)t), \partial S_\psi(Tx, q, t)) \\
&\geq c_0(1 - (1 - \epsilon))^n,
\end{aligned}
$$

and hence (ii) follows with $C = c_0$.   ∎

3.3.2   The engulfing property of the sections

The sections of a convex function whose Monge–Ampère measure satisfies (3.1.1) satisfy the following property similar to the one enjoyed by the Euclidean balls. Besides the importance of this property in the study of the linearized Monge–Ampère equation, it also permits us to establish that $\mathbb{R}^n$ equipped with the Monge–Ampère measure μ and the family of sections becomes a space of homogeneous type, see [AFT98].

Theorem 3.3.7 (engulfing property) *Assume that the Monge–Ampère measure μ associated with ϕ satisfies (3.1.1). Then there exists a constant $\theta > 1$ such that if $y \in S_\phi(x_0, p, t)$, then $S_\phi(x_0, p, t) \subset S_\phi(y, q, \theta t)$ for all $q \in \partial\phi(y)$.*

Proof. Let T be an affine transformation that normalizes the section $S_\phi(x_0, p, 2t)$, that is,

$$B(0, \alpha_n) \subset T(S_\phi(x_0, p, 2t)) \subset B(0, 1).$$

Let $\psi(y) = \phi(T^{-1}y), q_1 = (T^{-1})^t p$, and

$$\phi^*(y) = \psi(y) - \psi(Tx_0) - q_1 \cdot (y - Tx_0) - 2t.$$

By (3.2.3), $T(S_\phi(x_0, p, 2t)) = S_\psi(Tx_0, q_1, 2t)$. If $q_2 \in \partial\phi^*(Ty)$, then by Lemma 3.2.1

$$|q_2| \le \frac{2t}{\text{dist}(Ty, \partial S_\psi(Tx_0, q_1, 2t))}.$$

Since $y \in S_\phi(x_0, p, t)$, $Ty \in S_\psi(Tx_0, q_1, t)$. Thus, by Corollary 3.3.6(i)

$$|q_2| \le C_1 t.$$

By taking T^{-1}, the desired inclusion is equivalent to showing that

$$S_\psi(Tx_0, q_1, t) \subset S_\psi(Ty, (T^{-1})^t q, \theta t), \qquad \text{for all } q \in \partial\phi(y).$$

Let $z \in S_\psi(Tx_0, q_1, t)$. We want to show

$$\psi(z) < \psi(Ty) + (T^{-1})^t q \cdot (z - Ty) + \theta t, \qquad \text{for all } q \in \partial\phi(y).$$

We have $\partial\phi^* = \partial\psi - q_1$, and observe that $q \in \partial\phi(y)$ if and only if $(T^{-1})^t q \in \partial\psi(Ty)$. Hence, if $q \in \partial\phi(y)$, then $(T^{-1})^t q = q_2 + q_1$ with $q_2 \in \partial\phi^*(Ty)$. Therefore

$$
\begin{aligned}
&\psi(Ty) + (T^{-1})^t q \cdot (z - Ty) + \theta t \\
&= \psi(Ty) + q_2 \cdot (z - Ty) + q_1 \cdot (z - Ty) + \theta t \\
&\ge \psi(Tx_0) + q_1 \cdot (Ty - Tx_0) + q_2 \cdot (z - Ty) + q_1 \cdot (z - Ty) + \theta t \\
&= \psi(Tx_0) + q_1 \cdot (z - Tx_0) + q_2 \cdot (z - Ty) + \theta t \\
&\ge \psi(z) - t + q_2 \cdot (z - Ty) + \theta t = (*) \qquad \text{since } z \in S_\psi(Tx_0, q_1, t).
\end{aligned}
$$

Now $|q_2 \cdot (z - Ty)| \le C_1 t|z - Ty| \le 2C_1 t$. Hence $q_2 \cdot (z - Ty) \ge -2C_1 t$, and consequently

$$(*) \ge \psi(z) - t - 2C_1 t + \theta t = \psi(z) + (\theta - (2C_1 + 1))t.$$

The property now follows by picking $\theta \ge 2C_1 + 1$.   ∎

3.3.3 The size of normalized sections

The following theorem gives a quantitative estimate of the size of normalized sections. It says that if two sections intersect and we normalize the largest of them, then the other one looks like a ball with proportional radius at the scale in which the largest section is normalized. The statement below gives a more precise estimate than the one used in [CG96]; compare with condition (A) in that paper.

Theorem 3.3.8 *Assume that the Monge–Ampère measure μ associated with ϕ satisfies* (3.1.1). *There exist positive constants K_1, K_2, K_3 and ϵ such that if $S_\phi(z_0, p_0, r_0)$ and $S_\phi(z_1, p_1, r_1)$ are sections with $r_1 \le r_0$,*

$$S_\phi(z_0, p_0, r_0) \cap S_\phi(z_1, p_1, r_1) \ne \emptyset$$

and T is an affine transformation that normalizes $S_\phi(z_0, p_0, r_0)$, then

$$B\left(Tz_1, K_2 \frac{r_1}{r_0}\right) \subset T(S_\phi(z_1, p_1, r_1)) \subset B\left(Tz_1, K_1 \left(\frac{r_1}{r_0}\right)^\epsilon\right), \qquad (3.3.6)$$

and $Tz_1 \in B(0, K_3)$.

Proof. Let $\psi(y) = \dfrac{1}{r_0}\phi(T^{-1}y)$ and set

$$Tz_0 = x_0, \quad Tz_1 = x, \quad p = \frac{1}{r_0}(T^{-1})^t p_0, \quad q = \frac{1}{r_0}(T^{-1})^t p_1, \quad t = \frac{r_1}{r_0}.$$

Hence by (3.2.3) we have $T(S_\phi(z_0, p_0, r_0)) = S_\psi(x_0, p, 1)$ and

$$T(S_\phi(z_1, p_1, r_1)) = S_\psi(x, q, t).$$

For the rest of the proof, we shall omit the subscript ψ understanding that the defining function in all sections is ψ. Then the inclusions in the theorem are equivalent to

$$B(x, K_2 t) \subset S(x, q, t) \subset B(x, K_1 t^\epsilon).$$

Let us begin with the proof of the second inclusion. Since $S(x_0, p, 1)$ is normalized, the center of mass $c(S(x_0, p, 1)) = 0$. By Theorem 3.3.5, given $0 < \tau < 1$ there exists $0 < \lambda < 1$, $\lambda = \lambda(\tau, n) = 1 - \beta_n(1 - \tau)^n$ such that

$$S(x, q, \tau) \subset \lambda S(x, q, 1) = \{x_1 + \lambda(y - x_1) : y \in S(x, q, 1), x_1 = c(S(x, q, 1))\}$$
$$= \{(1 - \lambda)x_1 + \lambda y : y \in S(x, q, 1), x_1 = c(S(x, q, 1))\}.$$

In the same fashion

$$S(x, q, \tau^2) \subset \lambda S(x, q, \tau)$$
$$= \{(1 - \lambda)x_2 + \lambda y : y \in S(x, q, \tau), x_2 = c(S(x, q, \tau))\}$$
$$\subset \{(1 - \lambda)x_2 + \lambda(1 - \lambda)x_1 + \lambda^2 y : y \in S(x, q, 1),$$
$$x_1 = c(S(x, q, 1)), x_2 = c(S(x, q, \tau))\}.$$

For the same reason we obtain

$$S(x, q, \tau^{N+1}) \subset \lambda S(x, q, \tau^N),$$

and if we set $x_{i+1} = c(S(x, q, \tau^i))$, $i = 0, 1, 2, \ldots$, then continuing in this way we obtain

$$S(x, q, \tau^N) \subset \{(1 - \lambda) \sum_{i=0}^{N-1} \lambda^i x_{N-i} + \lambda^N y : y \in S(x, q, 1)\}.$$

If $x^* \in S(x_0, p, 1) \cap S(x, q, t)$, then Theorem 3.3.7 implies that $S(x, q, 1) \subset S(x^*, q', \theta)$ and $S(x_0, p, 1) \subset S(x^*, q', \theta)$ for all $q' \in \partial \psi(x^*)$. The last inclusion, again by the engulfing property, implies that $S(x^*, q', \theta) \subset S(x_0, p, \theta^2)$. On the other hand, the convexity of ψ implies that

$$S_\psi(x_0, p, r) \subset x_0 + r(S_\psi(x_0, p, 1) - x_0) = \{x_0 + r(z - x_0) : z \in S_\psi(x_0, p, 1)\}, \tag{3.3.7}$$

for each $r \geq 1$. Indeed, given $x \in S_\psi(x_0, p, r)$ we let $z = \frac{1}{r} x + (1 - \frac{1}{r}) x_0$ and we have $z \in S_\psi(x_0, p, 1)$. Hence

$$S(x, q, 1) \subset S(x_0, p, \theta^2) \subset x_0 + \theta^2(S(x_0, p, 1) - x_0).$$

Since $S(x_0, p, 1)$ is normalized,

$$\begin{aligned} x_0 &+ \theta^2(S(x_0, p, 1) - x_0) \\ &= (x_0 - \theta^2 x_0) + \theta^2 S(x_0, p, 1) \subset B(x_0 - \theta^2 x_0, \theta^2) \subset B(0, K), \end{aligned}$$

with $K = 2\theta^2 - 1$. Then $S(x, q, 1) \subset B(0, K)$ and consequently $x_{i+1} = c(S(x, q, \tau^i)) \in B(0, K)$, $i = 0, 1, \ldots$. Let $N \geq 0$ be such that $\tau^{N+1} < t \leq \tau^N$. Then

$$S(x, q, t) \subset S(x, q, \tau^N)$$

$$\subset \{y_N + \lambda^N y : y \in S(x, q, 1)\}, \qquad y_N = (1 - \lambda) \sum_{i=0}^{N-1} \lambda^i x_{N-i}$$

$$\subset B(y_N, \lambda^N K).$$

We have $N + 1 > \dfrac{\log_\lambda t}{\log_\lambda \tau}$, hence $\lambda^N < \lambda^{\frac{\log_\lambda t}{\log_\lambda \tau} - 1} = \dfrac{1}{\lambda} t^{\frac{\ln \lambda}{\ln \tau}}$. Since $|y_N| \leq (1 - \lambda) K \dfrac{1}{1 - \lambda} = K$, we obtain

$$S(x, q, t) \subset B\left(y_N, \frac{K}{\lambda} t^\epsilon\right) \subset B\left(x, 2\frac{K}{\lambda} t^\epsilon\right),$$

where $\epsilon = \dfrac{\ln \lambda}{\ln \tau}$.

Let us now show the first inclusion. If $y \in S(x_0, p, 1) \cap S(x, q, t)$, then by the engulfing property $S(x, q, t) \subset S(y, q', \theta t)$ and $S(x_0, p, 1) \subset S(y, q', \theta)$ for all $q' \in \partial \psi(y)$. Again by the engulfing property, Theorem 3.3.7, $S(y, q', \theta) \subset S(x_0, p, \theta^2)$ and consequently $S(x, q, t) \subset S(x_0, p, \theta^2)$, since $t \leq 1$. By (3.3.7)

$$S(x_0, p, 3\theta^2) \subset \{x_0 + 3\theta^2 (y - x_0) : y \in S(x_0, p, 1)\} \subset B(0, K),$$

with $K = 6\theta^2 - 1$. Let

$$\psi^*(z) = \psi(z) - \psi(x_0) - p \cdot (z - x_0) - 3\theta^2.$$

We claim that

$$\partial \psi^* \left(S(x_0, p, 2\theta^2) \right) \subset B(0, C\theta^2) \tag{3.3.8}$$

with a universal constant C. To show the claim, we first observe that if $\bar{\mu}$ is the Monge–Ampère measure associated with ψ and $S(x_0, p, 1)$ is normalized, then by Proposition 3.2.3 we have $\bar{\mu}(S(x_0, p, 1)) \approx 1$. Hence by the doubling property

$$\bar{\mu}(S(x_0, p, 2\theta^2)) \approx C(\theta), \qquad \bar{\mu}(S(x_0, p, 3\theta^2)) \approx C(\theta).$$

By Aleksandrov's estimate applied to $\psi_2(x) = \psi(x) - \psi(x_0) - p \cdot (x - x_0) - 2\theta^2$ on the section $S(x_0, p, 2\theta^2)$, we obtain

$$(\theta^2)^n \leq C \mathrm{dist}(S(x_0, p, \theta^2), \partial S(x_0, p, 2\theta^2)) \bar{\mu}(S(x_0, p, 2\theta^2)).$$

Thus

$$\mathrm{dist}(S(x_0, p, \theta^2), \partial S(x_0, p, 2\theta^2)) \geq C.$$

A similar argument yields

$$\mathrm{dist}(S(x_0, p, 2\theta^2), \partial S(x_0, p, 3\theta^2)) \geq C.$$

Hence by applying Lemma 3.2.1 to the function ψ^* in the set $S(x_0, p, 2\theta^2)$ we obtain (3.3.8). Let $x \in S(x_0, p, \theta^2)(\subset B(0, K))$. We shall pick K_2 such that $B(x, K_2 t) \subset S(x, q, t)$. Since $\mathrm{dist}(S(x_0, p, \theta^2), \partial S(x_0, p, 2\theta^2)) \geq C_1$ and $t \leq 1$, it follows that $B(x, C_1 t/4) \subset S(x_0, p, 2\theta^2)$. Let $y \in B(x, K_2 t)$ with $K_2 \leq C_1/4$. If $q' \in \partial \psi(y)$, then $\psi(x) \geq \psi(y) + q' \cdot (x - y)$ and $q' \in \partial \psi^*(y) + p$ by definition of ψ^*. From (3.3.8), $|q' - p| \leq C\theta^2$ and analogously $|q - p| \leq C\theta^2$. Thus $|q' - q| \leq 2C\theta^2$ and

$$0 \leq \psi(y) - \psi(x) - q \cdot (y - x) \leq -q' \cdot (x - y) - q \cdot (y - x)$$
$$\leq 2C\theta^2 |y - x| \leq 2C\theta^2 K_2 t < t,$$

by picking K_2 such that $2C\theta^2 K_2 < 1$. Thus $y \in S(x, q, t)$. The proof of Theorem 3.3.8 is now complete.   ∎

Remark 3.3.9 Theorem 3.3.8 implies that if the sections of the convex function ϕ are bounded sets and the corresponding Monge–Ampère measure satisfies (3.1.1), then ϕ is strictly convex. In fact, if $P_1 = (x_1, \phi(x_1))$ and $P_2 = (x_2, \phi(x_2))$ are points such that the segment $\overline{P_1 P_2}$ is contained in the graph of ϕ and $z_0 = t_0 x_1 + (1 - t_0)x_2$, $0 < t_0 < 1$, then any supporting hyperplane of ϕ at the point $(z_0, \phi(z_0))$ contains $\overline{P_1 P_2}$. Then $\overline{P_1 P_2} \subset S_\phi(z_0, p, t)$ for $p \in \partial\phi(z_0)$ and all $t > 0$. Then by Theorem 3.3.8 the segment $\overline{P_1 P_2}$ reduces to a point.

This implies that if the Monge–Ampère measure $M\phi$ satisfies the doubling property (3.1.1), then the sections $S_\phi(x_0, p, t)$ are strictly convex sets. Suppose by contradiction this is not true. Let $\ell_{x_0}(x)$ be the supporting hyperplane to u at x_0 defining the section $S_\phi(x_0, p, t)$. Then there exists $x_1 \neq x_2$ such that $\phi(x) - \ell_{x_0}(x) = t$ for x in the segment $\overline{x_1 x_2}$. Let $\psi(x) = \phi(x) - \ell_{x_0}(x)$. Then $M\psi$ is also doubling and therefore ψ must be strictly convex. But the segment $\overline{P_1 P_2}$ with $P_i = (x_i, t)$ is contained in the graph of ψ, contradiction.

Also as a consequence of Theorem 3.3.8 we obtain the following result of importance in the study of the solutions of the linearized Monge–Ampère equation, [CG97].

Theorem 3.3.10 *Assume that the Monge–Ampère measure μ associated with ϕ satisfies (3.1.1). Then*

(i) *There exist $C_0 > 0$ and $p_1 \geq 1$ such that for $0 < r < s \leq 1$, $t > 0$ and $x \in S_\phi(x_0, p, rt)$ we have*

$$S_\phi(x, q, C_0(s - r)^{p_1}t)) \subset S_\phi(x_0, p, st).$$

(ii) *There exist $C_1 > 0$ and $p_1 \geq 1$ such that for $0 < r < s < 1$, $t > 0$ and $x \in S_\phi(x_0, p, t) \setminus S_\phi(x_0, p, st)$ we have*

$$S_\phi(x, q, C_1(s - r)^{p_1}t) \cap S_\phi(x_0, p, rt) = \emptyset.$$

Proof. (i). Let T be an affine transformation normalizing $S_\phi(x_0, p, st)$. Then by (3.2.3)

$$T\left(S_\phi(x_0, p, st)\right) = S_\psi(Tx_0, q_1, s)$$

where $\psi(y) = \dfrac{1}{t}\phi(T^{-1}y)$, and $q_1 = \dfrac{1}{t}(T^{-1})^t p$. Also

$$T\left(S_\phi(x_0, p, rt)\right) = S_\psi(Tx_0, q_1, r),$$

$$T\left(S_\phi(x, q, C_0(s - r)^{p_1}t)\right) = S_\psi(Tx, q_2, C_0(s - r)^{p_1}),$$

where $q_2 = \dfrac{1}{t}(T^{-1})^t q$. To show (i), it suffices to prove that if $Tx \in S_\psi(Tx_0, q_1, r)$, then

$$S_\psi(Tx, q_2, C_0(s - r)^{p_1}) \subset S_\psi(Tx_0, q_1, s). \tag{3.3.9}$$

Set $\bar{r} = \dfrac{r}{s} < 1$. Let $\delta < s$ be chosen in a moment, and $z \in S_\psi(Tx, q_2, \delta)$. Then

$$
\begin{aligned}
\psi(z) &\le \psi(Tx) + q_2 \cdot (z - Tx) + \delta \\
&\le \psi(Tx_0) + q_1 \cdot (Tx - Tx_0) + r + q_2 \cdot (z - Tx) + \delta \\
&= \psi(Tx_0) + q_1 \cdot (z - Tx_0) + r + (q_2 - q_1) \cdot (z - Tx) + \delta.
\end{aligned}
$$

We have $q_1 \in \partial\psi(Tx_0)$, $q_2 \in \partial\psi(Tx)$, and $Tx, Tx_0 \in S_\psi(Tx_0, q_1, r)$. By applying Lemma 3.2.1 to the function $h(x) = \psi(x) - \psi(Tx_0) - q_1 \cdot (x - Tx_0) - s$ on the set $S_\psi(Tx_0, q_1, s)$, and using Corollary 3.3.6(i), it follows that

$$
(\partial\psi - q_1)(Tx) \subset B\left(0, \frac{-h(Tx)}{\operatorname{dist}(Tx, \partial S_\psi(Tx_0, q_1, s))}\right) \subset B\left(0, \frac{Cs}{(1 - \bar{r})^n}\right).
$$

This implies that $|q_2 - q_1| \le \dfrac{Cs}{(1 - \bar{r})^n} = \dfrac{Cs^{n+1}}{(s - r)^n}$. Therefore, by Theorem 3.3.8

$$
\begin{aligned}
\psi(z) &< \psi(Tx_0) + q_1 \cdot (z - Tx_0) + r + (q_2 - q_1) \cdot (z - Tx) + \delta \\
&\le \psi(Tx_0) + q_1 \cdot (z - Tx_0) + r + \frac{Cs^{n+1}}{(s - r)^n} K_1 \left(\frac{\delta}{s}\right)^\epsilon + \delta \\
&\le \psi(Tx_0) + q_1 \cdot (z - Tx_0) + r + \frac{CK_1}{(s - r)^n} \delta^\epsilon + \delta,
\end{aligned}
$$

since $\epsilon \le 1$. If $\delta = \left(\dfrac{(s - r)^{n+1}}{8CK_1}\right)^{1/\epsilon}$, then $\delta < (s - r)/2$ and (3.3.9) follows with

$$
C_0 = \left(\frac{1}{8K_1 C}\right)^{1/\epsilon} \quad \text{and} \quad p = \frac{n+1}{\epsilon}.
$$

(ii). Let T normalize $S_\phi(x_0, p, 2t)$ and $\psi(y) = \dfrac{1}{2t}\phi(T^{-1}y)$. Then

$$
T\left(S_\phi(x_0, p, 2t)\right) = S_\psi(Tx_0, q_1, 1),
$$

with $q_1 = \dfrac{1}{2t}(T^{-1})^t p$. It is sufficient to show that if $Tx \in S_\psi(Tx_0, q_1, 1/2) \setminus S_\psi(Tx_0, q_1, s/2)$, then

$$
S_\psi(Tx, q_2, C_1(s - r)^{p_1}/2) \cap S_\psi(Tx_0, q_1, r/2) = \emptyset.
$$

We have $q_1 \in \partial \psi (T x_0)$ and $q_2 \in \partial \psi (T x)$. By Corollary 3.3.6(i) and Lemma 3.2.1, $|q_2 - q_1| \leq C$. Let $\delta < 1$ and $z \in S_\psi (T x, q_2, \delta)$. By Theorem 3.3.8

$$
\begin{aligned}
\psi (z) &\geq \psi (T x) + q_2 \cdot (z - T x) \\
&\geq \psi (T x_0) + q_1 \cdot (T x - T x_0) + \frac{s}{2} + q_2 \cdot (z - T x) \\
&= \psi (T x_0) + q_1 \cdot (z - T x_0) + \frac{s}{2} + (q_2 - q_1) \cdot (z - T x) \\
&\geq \psi (T x_0) + q_1 \cdot (z - T x_0) + \frac{s}{2} - C K_1 \delta^\epsilon \\
&> \psi (T x_0) + q_1 \cdot (z - T x_0) + \frac{r}{2},
\end{aligned}
$$

if δ satisfies $\delta < \left(\dfrac{(s - r)}{2 C K_1} \right)^{p_1}$ and $p_1 = 1/\epsilon$. The desired conclusion then follows with $C_1 \leq \dfrac{1}{(2 C K_1)^{p_1}}$. ∎

From Theorem 3.3.10, we conclude that there exists $\delta > 0$ such that if $x \in S_\phi (x_0, p, 3t/4) \setminus S_\phi (x_0, p, t/2)$, then

$$
S_\phi (x, q, \delta t) \subset S_\phi (x_0, p, t) \setminus S_\phi (x_0, p, t/4).
$$

3.4 Notes

The notion of cross-section of solutions to the Monge–Ampère equation was introduced by L. A. Caffarelli in [Caf90a]. The exposition in this chapter largely follows the paper [GH00], and the geometric properties given in Theorems 3.3.7 and 3.3.8 are fundamental for the regularity theory developed in the following chapters. They are also used to study the linearizations, both elliptic and parabolic, of the Monge–Ampère equation, and in real harmonic analysis, see [Caf92], [Caf91], [CG96], [CG97] and [Hua99]. In particular, the covering lemma of Besicovitch type with cross-sections (Lemma 6.5.2) in Chapter 6 follows from these geometric properties. Condition (3.1.2) appears in [CG96] and is used there to establish with the aid of Lemma 6.5.2 a covering theorem of Calderón–Zygmund type; see [CG96, main Theorem] and our Theorem 6.3.3. This covering theorem is used in the study of the linearized Monge–Ampère operator, see [CG97]. Also, some geometric properties proved in this chapter imply that $\mathbb{R}^n$ equipped with the Monge–Ampère measure and the family of sections is a space of homogeneous type, see [AFT98].

Chapter 4

Convex Solutions of $\det D^2 u = 1$ in $\mathbb{R}^n$

4.1 Pogorelov's Lemma

We begin with an important and useful lemma due to Pogorelov.

Lemma 4.1.1 *Let $\Omega \subset \mathbb{R}^n$ be a convex open and bounded domain, and $u \in C^4(\Omega) \cap C^2(\bar{\Omega})$ a convex solution to the problem*

$$\det D^2 u = 1, \quad \text{in } \Omega,$$
$$u = 0, \quad \text{on } \partial\Omega.$$

Let α be a unit vector,

$$h(x) = |u(x)| D_{\alpha\alpha} u(x) e^{\frac{1}{2} D_\alpha u(x)^2},$$

and $M = \max_{\bar{\Omega}} h(x)$. Then there exists $P \in \Omega$ where the maximum M is attained and we have the inequality

$$M \leq C(n) \, e^{D_\alpha u(P)^2},$$

where $C(n)$ is a positive constant depending only on the dimension n.

Proof. Since $u = 0$ on $\partial\Omega$ and u is strictly convex in Ω, it follows that the maximum M is attained at some $P \in \Omega$. Since $D^2 u(P) > 0$, there exists a unimodular matrix O, i.e., $\det O = 1$, such that $O^t D^2 u(P) O$ is diagonal and if $\bar{u}(x) = u(Ox)$, then $D_1 \bar{u}(x) = D_\alpha u(Ox)$ and $D_{11} \bar{u}(x) = D_{\alpha\alpha} u(Ox)$; in particular, $D^2 \bar{u}(P')$ is diagonal where $P' = O^{-1} P$. To prove this statement, we first rotate the coordinates to have α as one of the axes. That is, let Q be an orthogonal matrix such that $Qe_1 = \alpha$, and first let $v(x) = u(Qx)$. Then the first column of Q is the vector α and we have $D_1v(x) = (D_\alpha u)(Qx)$ and $D_{11}v(x) = (D_{\alpha\alpha}u)(Qx)$. Next, given $A = (a_{ij})$, an $n \times n$ positive definite and symmetric matrix, consider

$$B = \begin{bmatrix} 1 & -\frac{a_{12}}{a_{11}} & -\frac{a_{13}}{a_{11}} & -\frac{a_{14}}{a_{11}} & \cdots & -\frac{a_{1n}}{a_{11}} \\ 0 & 1 & 0 & 0 & \cdots & 0 \\ 0 & 0 & 1 & 0 & \cdots & 0 \\ 0 & 0 & 0 & 1 & \cdots & 0 \\ \vdots & \vdots & \vdots & \vdots & \ddots & 0 \\ 0 & 0 & 0 & 0 & \cdots & 1 \end{bmatrix}.$$

We have

$$B^t A B = \begin{bmatrix} a_{11} & 0 \\ 0 & B_1 \end{bmatrix},$$

where B_1 is an $(n-1) \times (n-1)$ matrix. Since A is positive definite and symmetric, it follows that B_1 is also positive definite and symmetric. Hence there exists an orthogonal matrix O_1 such that $O_1^t B_1 O_1$ is diagonal. Let

$$\mathcal{O} = \begin{bmatrix} 1 & 0 \\ 0 & O_1 \end{bmatrix}.$$

Now, we choose $A = (D^2v)(Q^t P)$ and set $\bar{u}(x) = v(B\mathcal{O}x)$. Then $D^2\bar{u}((B\mathcal{O})^{-1}Q^t P)$ is diagonal. Combining the changes of coordinates, the matrix $O = QB\mathcal{O}$ does the job.

Therefore, we may assume that $\alpha = (1, 0, \ldots, 0)$ and so

$$h(x) = |u(x)|D_{11}u(x)e^{\frac{1}{2}(D_1u(x))^2},$$

and the matrix $D^2u(P)$ is diagonal.

Let L be the linearized operator at P,

$$L = \sum_{i=1}^{n} \frac{1}{u_{ii}(P)} D_{ii}.$$

Since h attains its maximum at P, it follows that the function

$$w = \log|u| + \log D_{11}u + \frac{1}{2}(D_1u)^2$$

also attains its maximum at P, and so $Dw(P) = 0$ and $D^2w(P) \leq 0$. Consequently

$$L(w)(P) \leq 0. \tag{4.1.1}$$

Now

$$w_i = \frac{u_i}{u} + \frac{u_{11i}}{u_{11}} + u_1 u_{1i},\qquad (4.1.2)$$

$$w_{ii} = \frac{u_{ii}}{u} - \frac{u_i^2}{u^2} + \frac{u_{11ii}}{u_{11}} - \frac{u_{11i}^2}{u_{11}^2} + u_{1i}^2 + u_1 u_{1ii}.$$

Hence

$$Lw(x) = \sum_{i=1}^{n} \frac{1}{u_{ii}(P)} D_{ii} w(x)$$

$$= \frac{1}{u(x)} \sum_{i=1}^{n} \frac{1}{u_{ii}(P)} u_{ii}(x) - \frac{1}{u(x)^2} \sum_{i=1}^{n} \frac{u_i(x)^2}{u_{ii}(P)}$$

$$+ \frac{1}{u_{11}(x)} L(u_{11})(x) - \frac{1}{u_{11}(x)^2} \sum_{i=1}^{n} \frac{u_{11i}(x)^2}{u_{ii}(P)}$$

$$+ \sum_{i=1}^{n} \frac{u_{1i}(x)^2}{u_{ii}(P)} + u_1(x) L(u_1)(x).$$

Since $D^2 u(P)$ is diagonal, differentiating $\det D^2 u = 1$ with respect to x_1 yields $L(u_1)(P) = 0$. In addition, $L(u)(P) = n$, and

$$L(u_{11})(P) = \sum_{k,l=1}^{n} \frac{u_{1kl}(P)^2}{u_{kk}(P) u_{ll}(P)}.$$

Then from (4.1.1)

$$Lw(P) = \frac{n}{u(P)} - \frac{1}{u(P)^2} \sum_{i=1}^{n} \frac{u_i(P)^2}{u_{ii}(P)} + \frac{1}{u_{11}(P)} \sum_{k,l=1}^{n} \frac{u_{1kl}(P)^2}{u_{kk}(P) u_{ll}(P)}$$

$$- \frac{1}{u_{11}(P)^2} \sum_{i=1}^{n} \frac{u_{11i}(P)^2}{u_{ii}(P)} + \sum_{i=1}^{n} \frac{u_{1i}(P)^2}{u_{ii}(P)} \le 0.$$

Now

$$\frac{1}{u_{11}(P)} \sum_{k,l=1}^{n} \frac{u_{1kl}(P)^2}{u_{kk}(P) u_{ll}(P)} - \frac{1}{u_{11}(P)^2} \sum_{i=1}^{n} \frac{u_{11i}(P)^2}{u_{ii}(P)}$$

$$= \frac{1}{u_{11}(P)} \left(\sum_{l=1}^{n} \frac{u_{11l}(P)^2}{u_{11}(P) u_{ll}(P)} + \sum_{k>1,l=1}^{n} \frac{u_{1kl}(P)^2}{u_{kk}(P) u_{ll}(P)} - \sum_{i=1}^{n} \frac{u_{11i}(P)^2}{u_{11}(P) u_{ii}(P)} \right)$$

$$= \frac{1}{u_{11}(P)} \sum_{k>1,l=1}^{n} \frac{u_{1kl}(P)^2}{u_{kk}(P) u_{ll}(P)}.$$

Hence

$$\frac{n}{u(P)} - \frac{1}{u(P)^2} \sum_{i=1}^n \frac{u_i(P)^2}{u_{ii}(P)} + \frac{1}{u_{11}(P)} \sum_{k>1, l=1}^n \frac{u_{1kl}(P)^2}{u_{kk}(P)\, u_{ll}(P)}$$

$$+ \sum_{i=1}^n \frac{u_{1i}(P)^2}{u_{ii}(P)} \leq 0. \qquad (4.1.3)$$

Since P is a maximum for w, from (4.1.2)

$$\frac{u_i(P)}{u(P)} + \frac{u_{11i}(P)}{u_{11}(P)} + u_1(P)\, u_{1i}(P) = 0, \quad i = 1, \cdots, n.$$

We have $u_{1i}(P) = 0$ for $i \neq 1$ since $D^2u(P)$ is diagonal. Then

$$\frac{u_i(P)}{u(P)} = -\frac{u_{11i}(P)}{u_{11}(P)}, \qquad \text{for } i \neq 1,$$

and so

$$\frac{u_i(P)^2}{u(P)^2} = \frac{u_{11i}(P)^2}{u_{11}(P)^2}, \qquad \text{for } i \neq 1.$$

Therefore

$$\frac{1}{u_{11}(P)} \sum_{k>1, l=1}^n \frac{u_{1kl}(P)^2}{u_{kk}(P)\, u_{ll}(P)}$$

$$= \sum_{k=2}^n \left(\frac{u_{1k1}(P)^2}{u_{kk}(P)\, u_{11}(P)^2} + \sum_{l=2}^n \frac{u_{1kl}(P)^2}{u_{11}(P)\, u_{kk}(P)\, u_{ll}(P)} \right)$$

$$= \sum_{k=2}^n \frac{u_{1k1}(P)^2}{u_{kk}(P)\, u_{11}(P)^2} + \text{positive terms}$$

$$= \sum_{k=2}^n \frac{u_k(P)^2}{u_{kk}(P)\, u(P)^2} + \text{positive terms}.$$

Inserting the last expression in (4.1.3) and dropping positive terms yields

$$\frac{n}{u(P)} - \frac{1}{u(P)^2} \frac{u_1(P)^2}{u_{11}(P)} + \sum_{i=1}^n \frac{u_{1i}(P)^2}{u_{ii}(P)} \leq 0,$$

which yields

$$\frac{n}{u(P)} - \frac{1}{u(P)^2} \frac{u_1(P)^2}{u_{11}(P)} + u_{11}(P) \leq 0.$$

Multiplying the last expression by $u_{11}(P)\, u(P)^2\, e^{u_1(P)^2}$ we obtain

$$h(P)^2 - n\, e^{u_1(P)^2/2}\, h(P) - u_1(P)^2\, e^{u_1(P)^2} \leq 0,$$

and so

$$h(P) \leq \frac{n\, e^{u_1(P)^2/2} + \sqrt{(n^2 + 4u_1(P)^2)e^{u_1(P)^2}}}{2}$$

$$= e^{u_1(P)^2/2} \left(\frac{n + \sqrt{n^2 + 4u_1(P)^2}}{2} \right) \leq C(n)\, e^{u_1(P)^2}.$$

∎

4.2 Interior Hölder estimates of D^2u

We use the notation

$$[u]_{\alpha, \Omega} = \sup_{x, y \in \Omega} \frac{|u(x) - u(y)|}{|x - y|^\alpha}.$$

Theorem 4.2.1 *Let $\Omega \subset \mathbb{R}^n$ be a convex open domain $B_{\alpha_n}(0) \subset \Omega \subset B_1(0)$ and $u \in C^4(\Omega) \cap C(\bar{\Omega})$ convex solution to the problem*

$$\det D^2u = 1, \quad in\ \Omega$$
$$u = 0, \quad on\ \partial\Omega.$$

Given $\epsilon > 0$ and $\Omega_\epsilon = \{x \in \Omega : u(x) < -\epsilon\}$ there exist constants $C = C(n)$ and $0 < \alpha < 1$ depending only on ϵ and the dimension n such that

$$[D^2u]_{\alpha, \Omega_\epsilon} \leq C.$$

Proof. **Step 1.** There exists a constant $C_0 > 0$, depending only on the structure, such that

$$\operatorname{dist}(\Omega_\epsilon, \partial\Omega) \geq C_0 \epsilon^n. \tag{4.2.1}$$

This follows from properties of the sections in Chapter 3. Indeed, let $m = \min_\Omega u$, and $x_0 \in \Omega$ such that $m = u(x_0)$. Since u is convex and $u = 0$ on $\partial\Omega$, it follows that $m < 0$ and $u(x) < 0$ for $x \in \Omega$. It is clear that $\ell(x) = m$ is a supporting hyperplane to u at $(x_0, u(x_0))$. Let $S(x_0, 0, t) = \{x : u(x) < m + t\}$. We have that $S(x_0, 0, -m) = \{x : u(x) < 0\} = \Omega$, and

$$S(x_0, 0, -m - \epsilon) = \Omega_\epsilon.$$

Since $S(x_0, 0, -m) = \Omega$ is normalized, by Corollary 3.3.6 we get that

$$\operatorname{dist}(S(x_0, 0, -m - \epsilon), \partial S(x_0, 0, -m))$$
$$= \operatorname{dist}(S(x_0, 0, \frac{-m - \epsilon}{-m}(-m)), \partial S(x_0, 0, -m))$$
$$\geq C_0 \left(1 - \frac{-m - \epsilon}{-m} \right)^n = C_0 \left(\frac{\epsilon}{-m} \right)^n.$$

By Proposition 3.2.3, $m \approx C_n$ and then (4.2.1) follows.

Step 2. We have the estimate

$$|Du(x)| \leq C\epsilon^{-n}, \qquad \text{for all } x \in \Omega_\epsilon. \tag{4.2.2}$$

By Lemma 3.2.1 we have

$$Du(x) \in B\left(0, \frac{\max_{\Omega_\epsilon} -u}{\text{dist}(\Omega_\epsilon, \partial\Omega)}\right), \qquad x \in \Omega_\epsilon.$$

Hence (4.2.2) follows from (4.2.1).

Step 3. If $|\alpha| = 1$, then

$$D_{\alpha\alpha} u(x) \leq C(\epsilon), \qquad \text{for all } x \in \Omega_{3\epsilon}. \tag{4.2.3}$$

In fact, consider the function $v(x) = u(x) + 2\epsilon$. We have

$$\det D^2 v = 1, \qquad \text{in } \Omega_{2\epsilon},$$
$$v = 0, \qquad \text{on } \partial\Omega_{2\epsilon}.$$

We apply Lemma 4.1.1 to v on the set $\Omega_{2\epsilon}$ and we obtain

$$\max_{\overline{\Omega_{2\epsilon}}} h(x) \leq C_n e^{D_\alpha u(P)^2},$$

where $h(x) = |v(x)| D_{\alpha\alpha} v(x) e^{\frac{1}{2} D_\alpha v(x)^2}$, and $h(P) = \max_{\overline{\Omega_{2\epsilon}}} h(x)$. Since $\Omega_{2\epsilon} \subset \Omega_\epsilon$, by (4.2.2) we get $|D_\alpha v(P)| = |D_\alpha u(P)| \leq C\epsilon^{-n}$, and consequently

$$h(x) = |v(x)| D_{\alpha\alpha} u(x) e^{\frac{1}{2} D_\alpha u(x)^2} \leq C_n e^{C\epsilon^{-2n}}, \qquad \text{for all } x \in \Omega_{2\epsilon}. \tag{4.2.4}$$

If $x \in \Omega_{3\epsilon}$, then $v(x) = u(x) + 2\epsilon < -3\epsilon + 2\epsilon = -\epsilon$, that is $|v(x)| > \epsilon$ in $\Omega_{3\epsilon}$ and from (4.2.4) we obtain

$$\epsilon D_{\alpha\alpha} u(x) \leq C_n e^{C\epsilon^{-2n}}, \qquad \text{for all } x \in \Omega_{3\epsilon}.$$

This yields (4.2.3) with $C(\epsilon) = C_n \dfrac{e^{C\epsilon^{-2n}}}{\epsilon}$.

Step 4. Let $\lambda_j(x)$, $j = 1, \ldots, n$ be the eigenvalues of $D^2 u(x)$. Then

$$\lambda_j(x) \geq \frac{1}{C(\epsilon)^{n-1}}, \qquad \text{for all } x \in \Omega_{3\epsilon}, \tag{4.2.5}$$

where $C(\epsilon)$ is the constant in Step 3.

In fact, since $\det D^2 u = 1$, we have $\lambda_1(x) \cdots \lambda_n(x) = 1$. Hence

$$\lambda_j(x) = \frac{1}{\lambda_1(x) \cdots \lambda_{j-1}(x)\lambda_{j+1}(x) \cdots \lambda_n(x)}.$$

From Step 3,

$$D_{\alpha\alpha}u(x) = \sum_{i,j=1}^{n} D_{ij}u(x)\alpha_i\alpha_j \leq C(\epsilon),$$

for all $\alpha = (\alpha_1, \ldots, \alpha_n)$; $|\alpha| = 1$. Since $D^2u(x)$ is symmetric, there exists an orthogonal matrix O such that

$$O D^2u(x)O^t = \begin{bmatrix} \lambda_1(x) & \cdots & 0 \\ 0 & \ddots & 0 \\ 0 & \cdots & \lambda_n(x) \end{bmatrix}.$$

If $z = (z_1, \ldots, z_n)$, $|z| = 1$, then $\langle D^2u(x)O^t z, O^t z \rangle = \langle O D^2u(x)O^t z, z \rangle = \sum_1^n \lambda_i(x)z_i^2$. Hence $\lambda_j(x) \leq C(\epsilon)$ for $1 \leq j \leq n$ and we obtain (4.2.5).

Step 5. Combining Steps 3 and 4 we obtain that

$$\frac{1}{C(\epsilon)^{n-1}} Id \leq D^2u(x) \leq C(\epsilon) Id, \qquad \text{for all } x \in \Omega_{3\epsilon}. \tag{4.2.6}$$

Step 6. We now recall the following result due to L. C. Evans, see [GT83, Section 17.4]:

Theorem 4.2.2 *Let $F \in C^2(\mathbb{R}^{n \times n})$, $g \in C^2(\Omega)$, $u \in C^4(\Omega)$ and*

$$F(D^2u) = g \qquad \text{in } \Omega.$$

Assume that

1. *F is uniformly elliptic with respect to u, i.e., there exist positive constants λ, Λ such that*

$$\lambda|\xi|^2 \leq F_{ij}\left(D^2u(x)\right)\xi_i\xi_j \leq \Lambda|\xi|^2,$$

for all $\xi \in \mathbb{R}^n$ and $x \in \Omega$. Here $F_{ij} = \dfrac{\partial F}{\partial a_{ij}}$.

2. *F is concave with respect to u, i.e., F is a concave function in the range of $D^2u(x)$, that is*

$$F_{ij,kl} = \frac{\partial^2 F}{\partial a_{ij} \partial a_{kl}}$$

satisfies

$$\frac{\partial^2 F}{\partial a_{ij} \partial a_{kl}}\left(D^2u(x)\right)u_{ij}(x)u_{kl}(x) \leq 0, \qquad \text{for all } x \in \Omega.$$

Then there exist positive constants C and $0 < \alpha < 1$ depending only on λ, Λ and n such that for each ball $B_{R_0} \subset \Omega$ and $R \leq R_0$ we have

$$\operatorname{osc}_{B_R} D^2u \leq C \left(\frac{R}{R_0}\right)^\alpha \left(\operatorname{osc}_{B_{R_0}} D^2u + R_0|Dg|_{0,\Omega} + R_0^2|D^2g|_{0,\Omega}\right).$$

We shall apply this result to the equation

$$F(D^2u) = \log\left(\det D^2u\right) = \log 1 = 0.$$

Here $\dfrac{\partial F}{\partial u_{ij}}(D^2u(x)) = u^{ij}(x)$, where $(u^{ij}(x))$ is the inverse matrix of $D^2u(x)$.

Since $D^2u(x)$ satisfies the inequalities in (4.2.6) and $\det D^2u = 1$, the inverse matrix also satisfies similar inequalities. Also, $F(D^2u)$ is concave. Since $g = 0$, we obtain

$$\operatorname{osc}_{B_R} D^2u \leq C \left(\frac{R}{R_0}\right)^\alpha \operatorname{osc}_{B_{R_0}} D^2u$$

for any ball $B_{R_0} \subset \Omega_{3\epsilon}$ and $R \leq R_0$. By Step 3, $\operatorname{osc}_{B_{R_0}} D^2u \leq C(\epsilon)$, for $B_{R_0} \subset \Omega_{3\epsilon}$. By covering $\Omega_{4\epsilon}$ with a finite number of balls contained in $\Omega_{3\epsilon}$ and by application of the previous inequality, we obtain the theorem. ∎

4.3 C^α estimates of D^2u

We can now prove the main result in this chapter.

Theorem 4.3.1 *Let $u \in C^4(\mathbb{R}^n)$ be a convex function such that $\det D^2u = 1$ in $\mathbb{R}^n$. There exist positive constants $C(n)$, C_1 and $0 < \alpha < 1$ such that for all $\lambda > 0$ we have the inequality*

$$[D_{ij}u]_{\alpha, B(0, C_1\lambda^{1/2})} \, \lambda^{\alpha/2} \leq C(n).$$

In particular, u must be a quadratic polynomial.

Proof. We may assume that

$$u(0) = 0; \qquad Du(0) = 0; \qquad \text{and } D^2u(0) = Id,$$

which implies that $u \geq 0$. Indeed, we first let $v(x) = u(x) - u(0) - Du(0) \cdot x$ and we have $v(0) = 0$, $Dv(0) = 0$ and $\det D^2v(x) = 1$. Since $D^2v(0)$ is symmetric and positive definite, there exists an orthogonal matrix O such that

$$O^t D^2v(0)O = \begin{bmatrix} d_1 & \cdots & 0 \\ \vdots & \ddots & \vdots \\ 0 & \cdots & d_n \end{bmatrix}$$

where $d_i > 0$, $i = 1, \ldots, n$. Let $w(x) = v(Ox)$. Then $D^2w(x) = O^t D^2 v(Ox)O$ and $\det D^2w = 1$ and also $w(0) = 0$, $Dw(0) = O^t(Dv)(O0) = O^t(Dv)(0) = 0$. Hence

$$D^2w(0) = \begin{bmatrix} d_1 & \cdots & 0 \\ \vdots & \ddots & \vdots \\ 0 & \cdots & d_n \end{bmatrix}.$$

Now let $\bar{w}(x) = w\left(\dfrac{x_1}{\sqrt{d_1}}, \ldots, \dfrac{x_n}{\sqrt{d_n}}\right)$. Then

$$D^2\bar{w}(x) =$$

$$\begin{bmatrix} \dfrac{1}{d_1} w_{11}\left(\dfrac{x_1}{\sqrt{d_1}}, \ldots, \dfrac{x_n}{\sqrt{d_n}}\right) & \cdots & \dfrac{1}{\sqrt{d_1}\sqrt{d_n}} w_{1n}\left(\dfrac{x_1}{\sqrt{d_1}}, \ldots, \dfrac{x_n}{\sqrt{d_n}}\right) \\ \vdots & \ddots & \vdots \\ \dfrac{1}{\sqrt{d_n}\sqrt{d_1}} w_{n1}\left(\dfrac{x_1}{\sqrt{d_1}}, \ldots, \dfrac{x_n}{\sqrt{d_n}}\right) & \cdots & \dfrac{1}{d_n} w_{nn}\left(\dfrac{x_1}{\sqrt{d_1}}, \ldots, \dfrac{x_n}{\sqrt{d_n}}\right) \end{bmatrix}$$

and so $\det D^2\bar{w}(x) = \dfrac{1}{d_1 \cdots d_n}(\det D^2w)\left(\dfrac{x_1}{\sqrt{d_1}}, \ldots, \dfrac{x_n}{\sqrt{d_n}}\right)$, and since $d_1 \cdots d_n = 1$, we get $\det D^2\bar{w}(x) = 1$ and $D^2\bar{w}(0) = Id$.

Let $\lambda > 0$ and $S_\lambda = \{x : u(x) < \lambda\}$ and E_λ be the ellipsoid of minimum volume containing S_λ with center at x_λ, the center of mass of S_λ. We note that if O is a rotation, then

$$O(S_\lambda) = \{z : u(O^{-1}z) < \lambda\}.$$

By changing u by $u(O^{-1}\cdot)$, we may assume that the axes of the ellipsoid E_λ lie on the coordinate axes. If $T = T_\lambda$ is an affine transformation that normalizes S_λ, that is $B_{\alpha_n}(0) \subset T(S_\lambda) \subset B_1(0)$, then $T(E_\lambda) = B_1(0)$ and $Tx = Ax + x_0$, $A = A_\lambda$ where A is a diagonal matrix

$$A = \begin{bmatrix} \mu_1 & \cdots & 0 \\ \vdots & \ddots & \vdots \\ 0 & \cdots & \mu_n \end{bmatrix}.$$

We also note that

$$E_\lambda = \left\{x : \frac{(x_1 - x_\lambda^1)^2}{a_1^2} + \cdots + \frac{(x_n - x_\lambda^n)^2}{a_n^2} \leq 1\right\},$$

and

$$Tx = \left(\frac{x_1 - x_\lambda^1}{a_1}, \ldots, \frac{x_n - x_\lambda^n}{a_n}\right),$$

that is

$$Tx = \begin{bmatrix} \dfrac{1}{a_1} & \cdots & 0 \\ \vdots & \ddots & \vdots \\ 0 & \cdots & \dfrac{1}{a_n} \end{bmatrix} (x - x_\lambda) = A(x - x_\lambda).$$

We claim that

$$\mu_i \approx \lambda^{-1/2}, \qquad i = 1, \cdots, n.$$

To show the claim, let

$$u^*(y) = \frac{1}{\gamma} u\left(T^{-1}y\right) - \frac{\lambda}{\gamma},$$

where γ is chosen such that $\det D^2u^*(y) = \dfrac{1}{\gamma^n} |\det T|^{-2} = 1$. Hence

$$|Du^*\left(T(S_\lambda)\right)| \approx 1.$$

The function u^* is convex and satisfies

$$\begin{aligned} \det D^2u^* &= 1, &&\text{in } T(S_\lambda), \\ u^* &= 0, &&\text{on } \partial T(S_\lambda). \end{aligned}$$

By Proposition 3.2.3,

$$|\min_{T(S_\lambda)} u^*| \approx C_n.$$

On the other hand, since $\max_{S_\lambda} u = \lambda$, we have that

$$|\min_{T(S_\lambda)} u^*| = \frac{\lambda}{\gamma}.$$

Hence

$$\frac{\lambda}{\gamma} \approx C_n. \tag{4.3.1}$$

We also note that $T^{-1}y = A^{-1}y + x_\lambda$. Applying Step 5 from Theorem 4.2.1 to u^* (see (4.2.6)) we obtain

$$C_2(\epsilon)Id \leq D^2u^*(x) \leq C_1(\epsilon)Id, \qquad \text{for all } x \in \{x : u^*(x) < -\epsilon\}.$$

Since $T = T_\lambda$ normalizes S_λ, by Theorem 3.3.8 applied to the sections S_λ, $S_{\tau\lambda}$ with $0 < \tau < 1$ we get that

$$B(T(0), K_2\tau) \subset T(S_{\tau\lambda}).$$

Also by (4.3.1), we obtain that

$$T(S_{\tau\lambda}) \subset \{x : u^*(x) < -(1-\tau)\beta\},$$

with $\beta > 0$ a constant depending only on n. Thus, there exists constants $\epsilon_0 > 0$ and $c_0 > 0$ such that

$$B(T(0), c_0) \subset \Omega^*_\epsilon = \{x : u^*(x) < -\epsilon\} \tag{4.3.2}$$

for all $\epsilon \leq \epsilon_0$. On the other hand,

$$D^2 u^*(y) = \frac{1}{\gamma}(A^{-1})^t D^2 u\left(A^{-1}y + x_\lambda\right) A^{-1},$$

and by letting $y = T(0) = -Ax_\lambda$ we obtain

$$D^2 u^*(T(0)) = \frac{1}{\gamma}\left(A^{-1}\right)^t D^2 u(0) A^{-1} = \frac{1}{\gamma}\left(A^{-1}\right)^t A^{-1}.$$

Consequently

$$C_2(\epsilon) Id \leq \frac{1}{\gamma}\left(A^{-1}\right)^t A^{-1} \leq C_1(\epsilon) Id.$$

Now

$$A^{-1} = \begin{bmatrix} \dfrac{1}{\mu_1} & \cdots & 0 \\ 0 & \ddots & 0 \\ 0 & \cdots & \dfrac{1}{\mu_1} \end{bmatrix},$$

and therefore

$$C_2 \leq \frac{1}{\gamma}\frac{1}{\mu_i^2} \leq C_1, \qquad i = 1, \dots, n.$$

Since $\dfrac{\lambda}{\gamma} \approx C_n$, the claim follows. Note also that since $|\det T||S_\lambda| \approx C$ we get that $|S_\lambda| \approx \lambda^{n/2}$.

We have

$$u^*(y) = \frac{1}{\gamma} u\left(\left(\frac{1}{\mu_1}y_1, \dots, \frac{1}{\mu_n}y_n\right) + x_\lambda\right) - \frac{\lambda}{\gamma},$$

and by Theorem 4.2.1

$$C(n) \geq [D_{ij}u^*]_{\alpha, \Omega^*_\epsilon}.$$

We now observe that if A is any invertible matrix and $v_A(x) = v(A^{-1}x + y_0)$, then

$$[v_A]_{\alpha, \Omega} \geq \frac{1}{\|A\|^\alpha}[v]_{\alpha, A^{-1}(\Omega)+y_0}.$$

We have

$$D_{ij}u^*(y) = \frac{1}{\gamma}\frac{1}{\mu_i\mu_j}D_{ij}u\left(\left(\frac{1}{\mu_1}y_1, \ldots, \frac{1}{\mu_n}y_n\right) + x_\lambda\right),$$

and so by (4.3.2)

$$C(n) \geq [D_{ij}u^*]_{\alpha,\Omega_\epsilon^*} \geq [D_{ij}u^*]_{\alpha,B(T(0),c_0)}$$

$$\geq \frac{1}{\gamma\mu_i\mu_j}\frac{1}{(\max_i \mu_i)^\alpha}[D_{ij}u]_{\alpha,A^{-1}(B(T(0),c_0))+x_\lambda}.$$

Since $T(0) = -Ax_\lambda$, it follows that $A^{-1}(B(T(0),c_0)) + x_\lambda = A^{-1}(B(0,c_0)) \approx B(0,\lambda^{1/2}c_0)$ and consequently,

$$C(n) \geq \lambda^{\alpha/2}[D_{ij}u]_{\alpha,B(0,\lambda^{1/2}c_0)}.$$

By letting $\lambda \to \infty$, we obtain that $D_{ij}u$ are constant on each bounded set and the proof is complete.   ∎

4.4   Notes

The characterization of global solutions to det $D^2u = 1$ given in Theorem 4.3.1, was done by Jörgens [Jör54] in one dimension (see also [Nit57]), by Calabi [Cal58] in dimensions less than six, and by Pogorelov [Pog72] in any dimension by extending Calabi's method; see also the paper by Cheng and Yau [CY86]. Recently, Caffarelli [Caf96] extended this characterization to viscosity solutions and in this chapter we have basically followed these ideas. Pogorelov's Lemma 4.1.1 appears in [Pog71]. For extensions of the results of this chapter to the parabolic case see [GH98].

Chapter 5

Regularity Theory for the Monge–Ampère Equation

5.1 Extremal points

Definition 5.1.1 *Let Ω be a convex subset of $\mathbb{R}^n$. The point $x_0 \in \partial\Omega$ is an extremal point of Ω if x_0 is not a convex combination of other points in $\bar{\Omega}$.*

Remark 5.1.2 *Let E be the set of extremal points of Ω. Then the convex hull of E equals $\bar{\Omega}$.*

Lemma 5.1.3 *Let $\Omega \neq \emptyset$ be a closed convex and bounded subset of $\mathbb{R}^n$. Then the set E of extremal points of Ω is nonempty.*

For a proof of the properties above see [Sch93, pp. 17–20].

Lemma 5.1.4 *Let x_0 be an extremal point of Ω. Then given $\delta > 0$ there exist a supporting hyperplane $\ell(x)$ at some point of $\partial\Omega$ (not necessarily x_0), and $\epsilon_0 > 0$ such that*

(a) $\Omega \subset \{x : \ell(x) \geq 0\}$,

(b) $\mathrm{diam}\{x \in \bar{\Omega} : 0 \leq \ell(x) \leq \epsilon_0\} < \delta$, and

(c) $0 \leq \ell(x_0) < \epsilon_0$.

Remark 5.1.5 To see that $\ell(x)$ can be a supporting hyperplane at a point other than x_0, consider for example in two dimensions a domain whose boundary is the segment on the x-axis joining the origin with the point $(-1, 0)$ and the arc of parabola $y = x^2$ for $x > 0$. Every supporting hyperplane at the point $x_0 = (0, 0)$ has slope zero.

Lemma 5.1.6 *Let Γ be a convex and bounded domain of $\mathbb{R}^n$, and u a convex function in Γ such that $u = 0$ on the boundary of Γ. If T is an affine transformation that normalizes Γ, then*

$$\{x \in \Gamma : dist(Tx, \partial T(\Gamma)) > \eta\} \subset \{x \in \Gamma : u(x) \le \eta\theta_n \min_\Gamma u\},$$

for all $0 < \eta < 1$, where θ_n is a constant depending only on n.

Proof.   Let x_0^* be the center of mass of Γ and $x_0 \in \Gamma$ such that $u(x_0) = \min_\Gamma u(x)$. Let T be an affine transformation that normalizes Γ. That is, $B_{\alpha_n}(0) \subset T(\Gamma) \subset B_1(0)$ and the center of mass of $T(\Gamma) = \Gamma^*$ is 0. Note that $v(x) = u(T^{-1}x)$ is convex in Γ^* and is zero on the boundary of Γ^*. Let $y_1 \in \Gamma^*$, $y_1 \ne 0$. Then $0 = \theta y_1 + (1 - \theta)\xi$ for some $\xi \in \partial\Gamma^*$ and $0 < \theta < 1$. Hence

$$\theta = \frac{|\xi|}{|y_1 - \xi|} \ge \frac{\alpha_n}{2} = \theta_n.$$

Thus

$$v(0) \le \theta v(y_1) + (1 - \theta)v(\xi) = \theta v(y_1) \le \theta_n v(y_1)$$

for all $y_1 \in \Gamma^*$. In particular, $v(0) \le \theta_n v(Tx_0) = \theta_n \min_\Gamma u$. Hence, if $z = 0 + \lambda(x - 0)$ with $x \in \Gamma^*$ and $0 < \lambda < 1$, then

$$v(z) \le \lambda v(x) + (1 - \lambda)v(0) \le (1 - \lambda)v(0) \le (1 - \lambda)\theta_n \min_\Gamma u.$$

That is

$$\lambda\Gamma^* \subset \{z \in \Gamma^* : v(z) \le (1 - \lambda)\theta_n \min_\Gamma u\}.$$

Given $0 < \eta < 1$, let us now prove that

$$\{z \in \Gamma^* : dist(z, \partial\Gamma^*) > \eta\} \subset (1 - \eta)\Gamma^*.$$

Indeed, for $z \in \Gamma^*$ we write $z = \lambda 0 + (1 - \lambda)x_\partial$, with $x_\partial \in \partial\Gamma^*$. Since the center of mass of Γ^* is zero and $dist(z, \partial\Gamma^*) > \eta$, it follows that

$$\lambda = \frac{|z - x_\partial|}{|x_\partial|} \ge \eta,$$

which yields $z \in (1 - \eta)\Gamma^*$. Therefore we obtain the inequality

$$\{z \in \Gamma^* : dist(z, \partial\Gamma^*) > \eta\} \subset \{z \in \Gamma^* : v(z) \le \eta\theta_n \min_\Gamma u\},$$

and the result follows by applying T^{-1}.   ∎

5.2 A result on extremal points of zeroes of solutions to Monge–Ampère

Theorem 5.2.1 *Let Ω be an open convex and bounded domain in $\mathbb{R}^n$, and u a convex function in Ω such that*

$$0 < \lambda \leq \det D^2 u \leq \Lambda$$

in the Aleksandrov sense. Assume that $u \geq 0$ in Ω and let

$$\Gamma = \{x \in \Omega : u(x) = 0\}.$$

If Γ is nonempty and contains more than one point, then Γ has no extremal points in the interior of Ω.

Proof. We proceed by contradiction. Assume that x_0 in the interior of Ω is an extremal point of Γ. Since $u \geq 0$, Γ is convex. Then by Lemma 5.1.4 given $\delta < \frac{1}{2}\mathrm{dist}(x_0, \partial\Omega)$, there exist $\epsilon_0 > 0$ and a supporting hyperplane $\ell(x)$ at some point $x_1 \in \partial\Gamma$ such that $\Gamma \subset \{x : \ell(x) \geq 0\}$, $\mathrm{diam}\{x \in \bar{\Gamma} : 0 \leq \ell(x) \leq \epsilon_0\} < \delta$ and $0 \leq \ell(x_0) < \epsilon_0$. Define

$$S = \{x \in \bar{\Gamma} : 0 \leq \ell(x) \leq \epsilon_0\},$$
$$\Pi_1 = \{x : \ell(x) = \epsilon_0\},$$
$$\Pi_2 = \{x : \ell(x) = 0\},$$
$$\Gamma_\epsilon = \{x \in \Omega : \ell(x) \leq \epsilon_0, u(x) \leq \epsilon(\epsilon_0 - \ell(x))\},$$

with $\epsilon > 0$. We have $S \subset \Gamma_\epsilon$ for all $\epsilon > 0$ and $\cap_{\epsilon>0}\Gamma_\epsilon = S \subset \mathrm{interior}(\Omega)$. Hence $\Gamma_\epsilon \subset \mathrm{interior}(\Omega)$ for ϵ sufficiently small. Note that Γ_ϵ is convex. Now move Π_2 in a parallel fashion to the left until it touches $\partial\Gamma_\epsilon$ at a point x_ϵ and let us denote by Π_3 the plane parallel to Π_2 at x_ϵ, i.e.,

$$\Pi_3 = \{x : \ell(x) = -\delta_\epsilon\}, \qquad x_\epsilon \in \Pi_3, \qquad \Gamma_\epsilon \subset \{x : \ell(x) \geq -\delta_\epsilon\},$$

with $\delta_\epsilon > 0$, see Figure 5.1.

Let $u_\epsilon(x) = u(x) - \epsilon(\epsilon_0 - \ell(x))$. We have that u_ϵ is convex and $\inf_{\Gamma_\epsilon} u_\epsilon < 0$. The first assertion is trivial, and the second follows because $x_1 \in \Gamma_\epsilon$ and $u_\epsilon(x_1) = u(x_1) - \epsilon(\epsilon_0 - \ell(x_1)) = 0 - \epsilon\epsilon_0 < 0$. We also have

$$\mathrm{dist}(\Pi_2, \Pi_3) = \delta_\epsilon \qquad \mathrm{dist}(\Pi_1, \Pi_2) = \epsilon_0,$$

so $\dfrac{\mathrm{dist}(\Pi_2, \Pi_3)}{\mathrm{dist}(\Pi_1, \Pi_2)} = \dfrac{\delta_\epsilon}{\epsilon_0}$ and $\delta_\epsilon \to 0$ as $\epsilon \to 0$.

Let us now study the quantity

$$\frac{|u_\epsilon(x_1)|}{|\inf_{\Gamma_\epsilon} u_\epsilon|}.$$

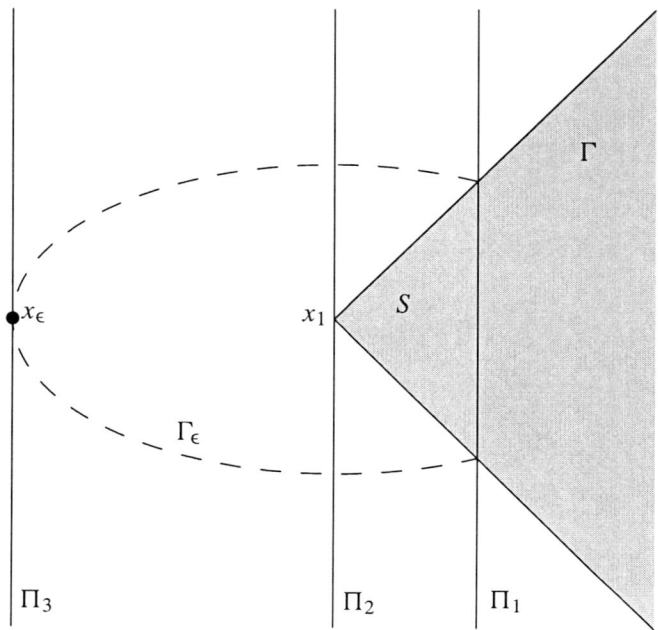

Figure 5.1. Theorem 5.2.1

We have $u(x) - \epsilon(\epsilon_0 - \ell(x)) \geq \left(\inf_{\Gamma_\epsilon} u\right) - \epsilon(\epsilon_0 - \ell(x))$ for all $x \in \Gamma_\epsilon$. Since $u \equiv 0$ on $\Pi_1 \cap \Gamma$, it follows that $\inf_{\Gamma_\epsilon} u = 0$. Hence $u_\epsilon(x) \geq -\epsilon(\epsilon_0 - \ell(x))$ for all $x \in \Gamma_\epsilon$, and so $\inf_{\Gamma_\epsilon} u_\epsilon \geq \inf_{\Gamma_\epsilon} \{-\epsilon(\epsilon_0 - \ell(x))\} = -\epsilon(\epsilon_0 + \delta_\epsilon)$. Consequently,

$$\frac{|u_\epsilon(x_1)|}{|\inf_{\Gamma_\epsilon} u_\epsilon|} = \frac{\epsilon \epsilon_0}{|\inf_{\Gamma_\epsilon} u_\epsilon|} \geq \frac{\epsilon_0}{\epsilon_0 + \delta_\epsilon},$$

and since $\delta_\epsilon \to 0$ as $\epsilon \to 0$, we obtain

$$\liminf_{\epsilon \to 0} \frac{|u_\epsilon(x_1)|}{|\inf_{\Gamma_\epsilon} u_\epsilon|} = 1.$$

Let T_ϵ be an affine transformation that normalizes Γ_ϵ, i.e., $B_{1/n}(0) \subset T_\epsilon(\Gamma_\epsilon) \subset B_1(0)$, and $u_\epsilon^*(x) = |\det T_\epsilon|^{2/n} u_\epsilon(T_\epsilon^{-1} x)$. We have $\lambda \leq \det D^2 u_\epsilon^* \leq \Lambda$ and $u_\epsilon^* = 0$ on $\partial \Gamma_\epsilon^*$, with $\Gamma_\epsilon^* = T_\epsilon(\Gamma_\epsilon)$. Then by the previous argument we get that

$$\frac{|u_\epsilon^*(T_\epsilon x_1)|}{|\inf_{\Gamma_\epsilon^*} u_\epsilon^*|} \geq C_1 > 0 \tag{5.2.1}$$

for ϵ small.

We shall show that

$$\operatorname{dist}(T_\epsilon x_1, \partial \Gamma_\epsilon^*) \to 0$$

as $\epsilon \to 0$. Let $\Pi_i^* = T_\epsilon \Pi_i$, with $i = 1, 2, 3$. We shall prove that

$$\frac{\text{dist}(\Pi_2^*, \Pi_3^*)}{\text{dist}(\Pi_1^*, \Pi_2^*)} \to 0, \tag{5.2.2}$$

as $\epsilon \to 0$. First observe that $\text{dist}(\Pi_1^*, \Pi_2^*) \leq C$, because $\text{dist}(\Pi_1^*, \Pi_2^*) \leq \text{dist}(\Pi_1^* \cap \Gamma_\epsilon^*, \Pi_2^* \cap \Gamma_\epsilon^*) \leq 2$, since $\Gamma_\epsilon^* \subset B_1(0)$. Also,

$$\frac{\text{dist}(\Pi_2^*, \Pi_3^*)}{\text{dist}(\Pi_1^*, \Pi_2^*)} = \frac{\text{dist}(\Pi_2, \Pi_3)}{\text{dist}(\Pi_1, \Pi_2)} = \frac{\delta_\epsilon}{\epsilon_0} \to 0$$

as $\epsilon \to 0$ and (5.2.2) follows, and consequently $\text{dist}(\Pi_2^*, \Pi_3^*) \to 0$. By Proposition 3.2.3 we have $|\inf_{\Gamma_\epsilon^*} u_\epsilon^*| \approx \left(Mu_\epsilon^*(\Gamma_\epsilon^*)\right)^{1/n}$, and from (5.2.1) we then obtain $|u_\epsilon^*(T_\epsilon x_1)| \geq C \left(Mu_\epsilon^*(\Gamma_\epsilon^*)\right)^{1/n}$. Let $\partial\Gamma_\epsilon^*$ be the part of the boundary of Γ_ϵ^* contained in the slab bounded by the planes Π_2^* and Π_3^* and let P_ϵ be a point on $\partial\Gamma_2^*$ such that the line through $T_\epsilon x_1$ and P_ϵ is perpendicular to Π_2^*. Then

$$\text{dist}(T_\epsilon x_1, \partial\Gamma_\epsilon^*) \leq \text{dist}(T_\epsilon x_1, \partial\Gamma_2^*) \leq |T_\epsilon x_1 - P_\epsilon| \leq \text{dist}(\Pi_2^*, \Pi_3^*),$$

which implies that $\text{dist}(T_\epsilon x_1, \partial\Gamma_\epsilon^*) \to 0$, as $\epsilon \to 0$. On the other hand, applying Aleksandrov's maximum principle, Theorem 1.4.2, to u_ϵ^* in Γ_ϵ^* we obtain

$$|u_\epsilon^*(T_\epsilon x_1)|^n \leq C_n \, \text{dist}(T_\epsilon x_1, \partial\Gamma_\epsilon^*) \, Mu_\epsilon^*(\Gamma_\epsilon^*).$$

Therefore, we get $\text{dist}(T_\epsilon x_1, \partial\Gamma_\epsilon^*) \geq C_n'$, a contradiction. ■

5.3 A strict convexity result

We begin with the following selection result.

Lemma 5.3.1 *Let Γ_j be a sequence of convex domains such that $B_{\alpha_n}(0) \subset \Gamma_j \subset B_1(0)$; and let u_j be a convex function in Γ_j that is a solution to*

$$\lambda \leq Mu_j \leq \Lambda \qquad \text{in } \Gamma_j,$$
$$u_j = 0 \qquad \text{on } \partial\Gamma_j.$$

Then there exist

(1) a normalized convex set Γ_∞,

(2) a convex function u_∞ that is a solution to $\lambda \leq Mu_\infty \leq \Lambda$ in Γ_∞, with $u_\infty|_{\partial\Gamma_\infty} = 0$,

and a subsequence of u_j converging to u_∞ uniformly on compact subsets of Γ_∞.

If in addition, for each j we have $x_j \in \Gamma_j$ such that $dist(x_j, \Gamma_j) \geq \epsilon$ and $\ell_j(x)$ a supporting hyperplane to u_j at x_j such that

$$S_j = \{x \in \Gamma_j : u_j(x) < \ell_j(x) + \frac{1}{j}\} \nsubseteq \{x \in \Gamma_j : u_j(x) < -C\epsilon\},$$

then there exist

(3) a point $x_\infty \in \Gamma_\infty$ such that $dist(x_\infty, \Gamma_\infty) \geq \epsilon$, and

(4) a supporting hyperplane ℓ_∞ to u_∞ at x_∞

such that

$$S_\infty = \{x \in \Gamma_\infty : u_\infty(x) \leq \ell_\infty(x)\} \nsubseteq T_\infty = \{x \in \Gamma_\infty : u_\infty(x) < -C\epsilon\}.$$

Proof.   There exists a convex function in $\mathbb{R}^n$, $z = F_j(x)$ such that

$$\Gamma_j = \{x \in \mathbb{R}^n : F_j(x) < 0\}$$

and $F_j(x) = 0$ for $x \in \partial\Gamma_j$. For example, $F_j(x)$ is the function that defines the cone in $\mathbb{R}^{n+1}$ generated by $\partial\Gamma_j \times \{0\}$ with vertex at $(0, -1)$.

We shall prove that there exists a subsequence of F_j, denoted also by F_j, such that

$$F_j(x) \rightarrow F(x)$$

for some F continuous in the Euclidean ball $B_2(0)$, where the convergence is uniform. Since Γ_j is normalized,

$$B_{\alpha_n}(0) \subset \Gamma_j \subset B_1(0).$$

If $p \in \partial F_j(x)$, then $|p| \leq c(n)$. We have $-1 \leq F_j(x) \leq c'(n)$ in $B_2(0)$. Therefore, the F_j are equicontinuous and bounded in $B_2(0)$ and by Arzèla–Ascoli, there is a subsequence that is uniformly convergent in $B_2(0)$.

Let $\Gamma_\infty = \{x \in B_2(0) : F(x) < 0\}$. The function F is convex and therefore so is Γ_∞. Let us show that

$$B_{1/n}(0) \subset \Gamma_\infty \subset B_1(0).$$

If $x \in \Gamma_\infty$, then $F(x) < 0$ and since $F_j(x) \rightarrow F(x)$, we have that $F_j(x) < 0$ for j large, that is $x \in \Gamma_j \subset B_1(0)$. We shall now prove that $B_{\alpha_n-\delta}(0) \subset \Gamma_\infty$ for every $\delta > 0$. Let $|x| < \alpha_n - \delta$. By similarity we have for $y \in \partial\Gamma_j$ that $\frac{|F_j(x)|}{|x - y|} = \frac{1}{|y|}$, that is, $|F_j(x)| = \frac{|x - y|}{|y|} \geq \frac{|y| - |x|}{|y|} \geq \delta$. That is $|F_j(x)| \geq \delta$, and so $F_j(x) \leq -\delta$. By letting $j \rightarrow \infty$ we get $F(x) \leq -\delta$, i.e., $x \in \Gamma_\infty$.

Next, we prove that for every compact $K \Subset \Gamma_\infty$ there exist positive constants $j_0(K)$ and $c(K)$ such that

$$K \subset \{x \in \Gamma_j : \mathrm{dist}(x, \partial \Gamma_j) > c(K)\},$$

for all $j \geq j_0(K)$. In fact, since $K \Subset \Gamma_\infty$, $\mathrm{dist}(K, \partial \Gamma_\infty) \geq \epsilon_0(K)$ and then $F(x) \leq -\eta_0(K)$ for all $x \in K$. Since $F_j \to F$, we have $F_j(x) \leq -\frac{\eta_0(K)}{2}$ for all $x \in K$ and so $K \subset \Gamma_j$. By the Aleksandrov maximum principle

$$|F_j(x)|^n \leq C_n \, \mathrm{dist}(x, \partial \Gamma_j) \, MF_j(\Gamma_j) \leq C_n' \, \mathrm{dist}(x, \partial \Gamma_j),$$

and therefore $\mathrm{dist}(x, \partial \Gamma_j) \geq \left(\frac{\eta_0(K)}{2} \right)^n$ for all $x \in K$, and the claim is proved.

We now claim that for every compact $K \Subset \Gamma_\infty$, there exists $C(K)$ such that

$$|u_j(x)| + |p| \leq C(K), \qquad \forall x \in K, \quad \forall p \in \partial u_j(x).$$

By the properties of normalized solutions, $|u_j(x)| \leq C(n, \lambda, \Lambda)$ in Γ_j. By Lemma 3.2.1, and the previous step, we have that $|p| \leq C(K)$ for every $p \in \partial u_j(x)$ for $x \in K$. Hence $|u_j(x) - u_j(z)| \leq C(K)|x - z|$ for $x, z \in K$. Then by Arzèla–Ascoli there exists a uniformly convergent subsequence in K. We now take a sequence of compacts $K_1 \subset K_2 \subset \cdots \subset \Gamma_\infty$ such that $\cup_j K_j = \Gamma_\infty$ and by a diagonal process we construct a subsequence $u_j(x)$ that is uniformly convergent in any compact $K \subset \Gamma_\infty$.

We define

$$u_\infty(x) = \lim_{j \to \infty} u_j(x), \qquad x \in \Gamma_\infty.$$

By Lemma 1.2.2, we have that u_∞ satisfies

$$\lambda \leq M u_\infty \leq \Lambda$$

in Γ_∞ in the generalized sense, and since $u_j(x) \leq 0$ we have that $u_\infty \leq 0$ in Γ_∞.

We shall prove that $u_\infty \in C\left(\overline{\Gamma_\infty}\right)$ and $u_\infty(x) = 0$ on $\partial \Gamma_\infty$. To this end, we first prove that for every $\eta > 0$ there exists a $j_0(\eta)$ such that

$$\{x \in \Gamma_j : \mathrm{dist}(x, \partial \Gamma_j) \geq \eta\} \subset \{x \in \Gamma_\infty : \mathrm{dist}(x, \partial \Gamma_\infty) \geq c\eta^n\}, \qquad (5.3.1)$$

for all $j \geq j_0(\eta)$. In fact, if $y_0 \in \Gamma_j$ and $\mathrm{dist}(y_0, \Gamma_j) \geq \eta$, then by Lemma 5.1.6 applied to F_j we get that $F_j(y_0) \leq -\theta_n \eta$. Since $F_j \to F$, we get that $F(y_0) \leq -C\eta/2$. By the Aleksandrov maximum principle applied to F (note that $F|_{\partial \Gamma_\infty} = 0$) we obtain

$$|F(y_0)|^n \leq C_n \mathrm{dist}(y_0, \partial \Gamma_\infty) \, MF(\Gamma_\infty),$$

which yields $\mathrm{dist}(y_0, \partial \Gamma_\infty) \geq \left(\frac{C\eta}{2} \right)^n$.

Next, we claim that

$$\{x \in \Gamma_\infty : u_\infty(x) < -\epsilon\} \Subset \Gamma_\infty,$$

for each $\epsilon > 0$. In fact, by Aleksandrov,

$$\{x \in \Gamma_j : u_j(x) < -\epsilon/2\} \subset \{x \in \Gamma_j : \text{dist}(x, \partial\Gamma_j) \geq C\epsilon^n\},$$

and hence by (5.3.1)

$$\{x \in \Gamma_j : \text{dist}(x, \partial\Gamma_j) \geq C\epsilon^n\} \subset \{x \in \Gamma_\infty : \text{dist}(x, \partial\Gamma_\infty) \geq C\epsilon^{n^2}\} \stackrel{\text{def}}{=} K(\epsilon),$$

for $j \geq j_0(\epsilon)$. Therefore

$$\cup_{j \geq j_0}\{x \in \Gamma_j : u_j(x) < -\epsilon/2\} \subset K(\epsilon). \tag{5.3.2}$$

Since $u_j \to u_\infty$, it follows that

$$\{x \in \Gamma_\infty : u_\infty(x) \leq -\epsilon\} \subset K(\epsilon).$$

It is now clear from the definition of $K(\epsilon)$ that $\lim_{x \to \partial\Gamma_\infty} u_\infty(x) = 0$.

It remains to construct the point x_∞ and the supporting plane ℓ_∞ with the desired properties. We have that

$$S_j = \{x \in \Gamma_j : u_j(x) < \ell_j(x) + \frac{1}{j}\} \not\subset \{x \in \Gamma_j : u_j(x) < -C\epsilon\} = T_j,$$

and consequently a point $y_j \in \Gamma_j$ such that $u_j(y_j) \leq \ell_j(y_j) + \frac{1}{j}$, with ℓ_j a supporting hyperplane to u_j at x_j and $u_j(y_j) \geq -C\epsilon$ and $\text{dist}(x_j, \partial\Gamma_j) \geq \epsilon$. By choosing the constant C appropriately and depending only on the structure, we may assume that $u_j(y_j) = -C\epsilon$. Indeed, by Lemma 5.1.6, $u_j(x_j) \leq \epsilon\,\theta_n\,\min_{\Gamma_j} u_j \leq -\epsilon\,C(n, \lambda, \Lambda)$. By joining the points x_j and y_j with a segment, by continuity we can pick $\bar{y}_j$ on this segment so that $u_j(\bar{y}_j) = -C\epsilon$.

By (5.3.1), $x_j \in \Gamma_\infty$ and

$$\text{dist}(x_j, \partial\Gamma_\infty) \geq C\epsilon^n, \tag{5.3.3}$$

for j large. Also by Aleksandrov,

$$(C\epsilon)^n = |u_j(y_j)|^n \leq C_n\,\text{dist}(y_j, \partial\Gamma_j)\,Mu_j(\Gamma_j),$$

which implies $\text{dist}(y_j, \partial\Gamma_j) \geq C'\epsilon^{1/n}$. So again by (5.3.1)

$$\text{dist}(y_j, \partial\Gamma_\infty) \geq C(\epsilon, n, \Lambda).$$

Then by choosing a subsequence $x_j \to x_\infty$ and $y_j \to y_\infty$. Let $\ell_j(x) = u_j(x_j) + p_j \cdot (x - x_j)$. By Lemma 3.2.1, we have $|p_j| \leq C_\epsilon$ for j large and by choosing a subsequence, $p_j \to p_\infty$. Recall that $u_j \to u_\infty$ uniformly on compact subsets of Γ_∞. By (5.3.3) we have $\text{dist}(x_\infty, \partial \Gamma_\infty) \geq C\epsilon^n$. Now

$$u_j(x) \geq u_j(x_j) + p_j \cdot (x - x_j), \qquad \forall x \in \Gamma_j,$$

and by letting $j \to \infty$,

$$u_\infty(x) \geq u_\infty(x_\infty) + p_\infty \cdot (x - x_\infty), \qquad \forall x \in \Gamma_\infty.$$

Then $\ell_\infty(x) = u_\infty(x_\infty) + p_\infty \cdot (x - x_\infty)$ is a supporting hyperplane to u_∞ at x_∞. Also, $u_j(y_j) = -C\epsilon$, which implies $u_\infty(y_\infty) = -C\epsilon$, and also $y_\infty \in \Gamma_\infty$. In addition, $u_j(y_j) \leq \ell_j(y_j) + \frac{1}{j}$ and by letting $j \to \infty$ we get $u_\infty(y_\infty) \leq \ell_\infty(y_\infty)$, which means that $y_\infty \in S_\infty$ and $y_\infty \notin T_\infty$. This completes the proof of the lemma. ∎

Lemma 5.3.2 *Let u be convex in Γ, not necessarily normalized, such that for some $x_0 \in \Gamma^\circ$ we have*

$$u(x) > u(x_0) \qquad \text{for all } x \in \partial \Gamma.$$

Let ℓ_{x_0} be a supporting hyperplane to u at x_0. If the set

$$E = \{x \in \Gamma : u(x) = \ell_{x_0}(x)\}$$

has more than one point, then E has an extremal point in the interior of Γ.

Proof. Suppose by contradiction that E has no extremal points in Γ°. If E^* is the set of extremal points of E, then $E^* \subset \partial \Gamma$. By Remark 5.1.2, $x_0 = \sum_{i=1}^N \lambda_i x_i$ for some $x_i \in E^*$ and $\lambda_i \geq 0$ with $\sum_{i=1}^N \lambda_i = 1$. Then

$$u(x_0) = \ell_{x_0}(x_0) = \sum_{i=1}^N \lambda_i \ell_{x_0}(x_i) = \sum_{i=1}^N \lambda_i u(x_i)$$

$$> u(x_0) \sum_{i=1}^N \lambda_i = u(x_0),$$

a contradiction. ∎

We are now in a position to prove the main result of this section.

Theorem 5.3.3 *Given Γ a convex and bounded normalized domain in $\mathbb{R}^n$, consider u convex in Γ, a generalized solution of the problem*

$$\lambda \leq \det D^2 u \leq \Lambda, \qquad u|_{\partial \Gamma} = 0. \tag{5.3.4}$$

Then for each $\epsilon > 0$ there exists $\delta = \delta(\epsilon)$ such that for all Γ normalized, for all x_0 such that $\mathrm{dist}(x_0, \partial\Gamma) \geq \epsilon$, for all u that are solutions of (5.3.4), and for all $\ell(x)$ supporting hyperplanes of u at x_0, we have

$$\{x \in \Gamma : u(x) < \ell(x) + \delta\} \Subset \Gamma$$

where the inclusion is compact. Moreover

$$\{x \in \Gamma : u(x) < \ell(x) + \delta\} \subset \{x \in \Gamma : u(x) < -C\epsilon\},$$

with $C = C(n, \lambda, \Lambda)$. We remark that δ is independent of x_0, u, ℓ and Γ, and depends only on ϵ, λ, Λ and n.

Proof. The proof is by contradiction. Suppose that there exists $\epsilon > 0$ such that for each $\delta = \frac{1}{j}$ there exist a point x_j, a normalized convex set Γ_j, $\mathrm{dist}(x_j, \Gamma_j) \geq \epsilon$, a solution u_j to (5.3.4) in Γ_j and a supporting hyperplane ℓ_j to u_j at x_j such that

$$S_j = \{x \in \Gamma_j : u_j(x) < \ell_j(x) + \frac{1}{j}\} \nsubseteq \{x \in \Gamma_j : u_j(x) < -C\epsilon\}.$$

By Lemma 5.3.1 we have

> a normalized convex set Γ_∞,
>
> a convex function u_∞, a solution of (5.3.4) in Γ_∞,
>
> a point $x_\infty \in \Gamma_\infty$ such that $\mathrm{dist}(x_\infty, \Gamma_\infty) \geq \epsilon$,
>
> and a supporting hyperplane ℓ_∞ to u_∞ at x_∞

such that

$$S_\infty = \{x \in \Gamma_\infty : u_\infty(x) \leq \ell_\infty(x)\} \nsubseteq T_\infty = \{x \in \Gamma_\infty : u_\infty(x) < -C\epsilon\}.$$

That is, there exists $z \in S_\infty$ such that $u_\infty(z) \geq -C\epsilon$. By Lemma 5.1.6, $x_\infty \in T_\infty$, i.e., $u_\infty(x_\infty) < -C\epsilon$, and so the segment $\overline{x_\infty z}$ has more than one point. We have $\overline{x_\infty z} \subset S_\infty = \{x \in \Gamma_\infty : u_\infty(x) = \ell_\infty(x)\}$, and by Theorem 5.2.1, S_∞ has no extremal points in the interior of Γ_∞. On the other hand, by Lemma 5.3.1(2) we have that $u_\infty(x_\infty) < u_\infty(x) = 0$ for all $x \in \partial\Gamma_\infty$ and from Lemma 5.3.2 we conclude that S_∞ must have an extremal point in the interior of Γ_∞, a contradiction. This completes the proof of the theorem. ∎

5.4 $C^{1,\alpha}$ regularity

Lemma 5.4.1 *Let Γ be a convex normalized domain and u a generalized solution to $\lambda \leq Mu \leq \Lambda$ in Γ with $u|_{\partial\Gamma} = 0$. Given $0 < \eta \leq 1$, set*

$$\Gamma_\eta = \{x \in \Gamma : u(x) < (1 - \eta) \min_\Gamma u\},$$

and $u(x_0) = \min_\Gamma u$. Then there exists $\mu = \mu(\lambda, \Lambda, n) < 1$ such that

$$\frac{1}{2}\Gamma \subset \mu\,\Gamma_{1/2}, \qquad\qquad (5.4.1)$$

where the dilations are with respect to x_0.

Proof. Notice that by Theorem 5.2.1, there can be only one point where the minimum is attained.

Suppose by contradiction that for each $j = 1, 2, \ldots$ and each $\mu_j = 1 - \dfrac{1}{j}$, there exists a normalized domain Γ_j and a solution u_j of $\lambda \le Mu_j \le \Lambda$ in Γ_j with $u_j|_{\partial\Gamma} = 0$, such that $\frac{1}{2}\Gamma_j \not\subset \mu_j\,(\Gamma_j)_{1/2}$, where the dilation is with respect to x_j and $\min_{\Gamma_j} u_j = u_j(x_j)$. Notice that $x_j \in \frac{1}{2}\Gamma_j \cap \mu_j\,(\Gamma_j)_{1/2}$; thus $\frac{1}{2}\Gamma_j \cap \partial\left(\mu_j\,(\Gamma_j)_{1/2}\right) \ne \emptyset$ and so we can pick $y_j \in \frac{1}{2}\Gamma_j \cap \partial\left(\mu_j\,(\Gamma_j)_{1/2}\right)$. There is a subsequence $y_j \to y_\infty$. By the selection Lemma 5.3.1 there is a normalized domain Γ_∞ and a convex function u_∞, a solution to $\lambda \le Mu_\infty \le \Lambda$ in Γ_∞, with $u_\infty|_{\Gamma_\infty} = 0$, and $u_j \to u_\infty$, for some subsequence, uniformly on compact subsets of Γ_∞. We claim that

$$y_\infty \in \frac{1}{2}\overline{\Gamma_\infty} \cap \partial\,(\Gamma_\infty)_{1/2}\,, \qquad\qquad (5.4.2)$$

where the dilation is with respect to x_∞, $u_\infty(x_\infty) = \min_{\Gamma_\infty} u_\infty$. Assume (5.4.2) for now. Construct the line segment through x_∞ and y_∞, crossing $\partial\Gamma_\infty$ at y_∞^*. Since $y_\infty \in \partial\,(\Gamma_\infty)_{1/2}$, it follows that $u_\infty(y_\infty) = \frac{1}{2}\min_{\Gamma_\infty} u_\infty$, and $u_\infty(y_\infty^*) = 0$ since $y_\infty^* \in \partial\Gamma_\infty$. We have $y_\infty = \theta\,x_\infty + (1-\theta)\,y_\infty^*$ for some $\theta \in (0, 1)$. By convexity

$$u_\infty(y_\infty) = \frac{1}{2}\min_{\Gamma_\infty} u_\infty \le \theta\,u_\infty(x_\infty) + (1-\theta)\,u_\infty(y_\infty^*) = \theta\min_{\Gamma_\infty} u_\infty,$$

and since $\min_{\Gamma_\infty} u_\infty < 0$, we get $\theta \le 1/2$. Suppose $\theta < 1/2$. Then $y_\infty = x_\infty + (1-\theta)\,(y_\infty^* - x_\infty) \in (1-\theta)\,\partial\Gamma_\infty$, but this is impossible since $y_\infty \in \frac{1}{2}\overline{\Gamma_\infty}$. Therefore, $\theta = 1/2$ and so y_∞ is the midpoint of the segment $\overline{x_\infty y_\infty^*}$. This implies that u_∞ is affine on this segment and hence if ℓ is a supporting hyperplane containing this segment, we have that the set $E = \{u_\infty = \ell\}$ has more than one point. By Theorem 5.2.1 applied to $u_\infty - \ell$, it follows that E has no extremal points in the interior of Γ_∞. On the other hand, $u(x) > u(x_\infty)$ for all $x \in \partial\Gamma_\infty$ and by Lemma 5.3.2, E has an extremal point in the interior of Γ_∞. This is a contradiction.

Let us prove (5.4.2). We have $y_j = x_j + \dfrac{1}{2}\,(z_j - x_j)$ with $z_j \in \Gamma_j$. By Aleksandrov's estimate, Theorem 1.4.2, $\text{dist}(x_j, \partial\Gamma_j) \ge \epsilon_0$ since $u_j(x_j) \approx C(n, \lambda, \Lambda)$. By (5.3.1), $x_j \in \Gamma_\infty$ and $\text{dist}(x_j, \partial\Gamma_\infty) \ge C\,\epsilon_0^n$, for j large. By passing to a subsequence, $x_j \to \bar{x}$, and since $\text{dist}(\bar{x}, \partial\Gamma_\infty) \ge C\,\epsilon_0^n$, there exists $\epsilon > 0$ such that $B_\epsilon(\bar{x}) \subset \Gamma_\infty$. Since $u_j \to u_\infty$ uniformly on compact subsets, $u_j(x_j) \to u_\infty(\bar{x})$.

Also $u_j(x_j) \leq u_j(x)$ for $x \in \Gamma_j$, and letting $j \to \infty$ we get $u_\infty(\bar{x}) \leq u_\infty(x)$ in Γ_∞. Therefore $\bar{x} = x_\infty$. Now select a subsequence so that $z_j \to z_\infty$. We have $F_j(z_j) < 0$, where F_j are the defining cones of Γ_j in the proof of the selection Lemma 5.3.1. Hence $F_\infty(z_\infty) \leq 0$, so $z_\infty \in \overline{\Gamma_\infty}$. Then $y_\infty = x_\infty + \frac{1}{2}(z_\infty - x_\infty)$, so $y_\infty \in \frac{1}{2}\overline{\Gamma_\infty}$. It remains to show that $y_\infty \in \partial(\Gamma_\infty)_{1/2}$. We have $y_j \in \partial\left(\mu_j\,(\Gamma_j)_{1/2}\right) = \mu_j\,\partial(\Gamma_j)_{1/2}$, so $y_j = x_j + \mu_j\,(z_j - x_j)$, $z_j \in \partial(\Gamma_j)_{1/2}$, $u_j(z_j) = \frac{1}{2}\min_{\Gamma_j} u_j$. Select a subsequence $z_j \to z^*$ (again Aleksandrov guarantees that the z_j are away from the boundary), and by uniform convergence $u_\infty(z^*) = \lim_{j\to\infty} u_j(z_j)$. Hence $u_\infty(z^*) = \frac{1}{2}u_\infty(x_\infty)$, i.e., $z^* \in \partial(\Gamma_\infty)_{1/2}$. Passing to the limit in $y_j = x_j + \mu_j\,(z_j - x_j)$ yields $y_\infty = z^*$.

This completes the proof of the lemma. ∎

Remark 5.4.2 Given $0 < \eta \leq 1$, let $V_\eta(x)$ be the function whose graph is the cone in $\mathbb{R}^{n+1}$ with vertex at $(x_0, 0)$ and passing through the set $\{x \in \Gamma : u(x) = (1 - \eta)\,u(x_0)\} \times \{-\eta\,u(x_0)\}$. Then (5.4.1) is equivalent to $V_{1/2}(x) \leq \mu\,V_1(x)$ for all $x \in \Gamma$. Indeed, given $x \in \Gamma$ consider the ray $\overrightarrow{\ell}$ emanating from x_0 and passing through x, and let $x' = \overrightarrow{\ell} \cap \partial\Gamma_{1/2}$ and $x'' = \overrightarrow{\ell} \cap \partial\Gamma$. By similarity of triangles we get that $V_{1/2}(x) = -\dfrac{u(x_0)}{2}\,\dfrac{|x - x_0|}{|x' - x_0|}$, and $V_1(x) = -u(x_0)\,\dfrac{|x - x_0|}{|x'' - x_0|}$. Since $x_0 + \dfrac{1}{2}(x'' - x_0) \in \dfrac{1}{2}\Gamma$, by (5.4.1) there exists $z \in \Gamma_{1/2}$ such that $\frac{1}{2}(x'' - x_0) = \mu\,(z - x_0)$ and so $|x'' - x_0| = 2\mu|z - x_0| \leq 2\mu|x' - x_0|$, and we get $V_{1/2}(x) \leq \mu\,V_1(x)$. Conversely, if $x \in \Gamma$ then $\dfrac{1}{2}\dfrac{|x - x_0|}{|x' - x_0|} \leq \mu\,\dfrac{|x - x_0|}{|x'' - x_0|}$. So $\dfrac{1}{2}|x - x_0| \leq \mu\,\dfrac{|x - x_0|}{|x'' - x_0|}\,|x' - x_0| \leq \mu\,|x' - x_0|$. That is, $x_0 + \dfrac{1}{2}(x - x_0) \in \mu\,\Gamma_{1/2}$.

We now show that the constant μ in Lemma 5.4.1 is invariant under affine changes of variables.

Lemma 5.4.3 *Let Γ be a bounded convex domain (not necessarily normalized), and u a generalized solution to $\lambda \leq Mu \leq \Lambda$ in Γ with $u|_{\partial\Gamma} = \beta$, $\beta \in \mathbb{R}$. Suppose $u(x_0) = \min_\Gamma u$. Then*

$$\frac{1}{2}\Gamma \subset \mu\,\left\{x \in \Gamma : u(x) \leq \frac{1}{2}\left(\min_\Gamma u + \beta\right)\right\},$$

where the dilation is with respect to x_0, and $0 < \mu < 1$ is the constant in Lemma 5.4.1.

Proof. Let T be an affine transformation normalizing Γ, $T(\Gamma) = \Gamma^*$, and $v(x) = |\det T|^{2/n}\left(u(T^{-1}x) - \beta\right)$. The function v attains its minimum at Tx_0. Applying Lemma 5.4.1 to v in Γ^* yields $\frac{1}{2}\Gamma^* \subset \mu\,(\Gamma^*)_{1/2}$, with dilations with respect to Tx_0. If $0 < \eta < 1$, then $(\Gamma^*)_\eta = T(\{x \in \Gamma : u(x) < (1 - \eta)\min_\Gamma u + \beta\,\eta\})$.

Letting $\eta = 1/2$ and $E = \{x \in \Gamma : u(x) < \frac{1}{2}(\min_\Gamma u + \beta)\}$, we get $(\Gamma^*)_{1/2} = T(E)$. On the other hand, $\mu\,(\Gamma^*)_{1/2} = T(\mu\,E)$ and $\frac{1}{2}\Gamma^* = T(\frac{1}{2}\,\Gamma)$, where in both identities the first dilation is with respect to Tx_0 and the second with respect to x_0. The lemma then follows by taking T^{-1}. ∎

Corollary 5.4.4 *Under the hypotheses of Lemma 5.4.1, we have that*

$$\Gamma \subset (2\,\mu)^k\, \Gamma_{1/2^k}, \qquad for\ k = 1, 2, \ldots.$$

Proof. By Lemma 5.4.1, $\Gamma \subset 2\mu\,\Gamma_{1/2}$. Applying Lemma 5.4.3 to $\Gamma_{1/2}$ with $\beta = \frac{1}{2}\min_\Gamma u$, and noticing that $\min_{\Gamma_{1/2}} u = u(x_0)$, we have that

$$\Gamma_{1/2} \subset 2\mu\,\{x \in \Gamma_{1/2} : u(x) < \frac{1}{2}(\min_\Gamma u + \frac{1}{2}\min_\Gamma u)\}.$$

So

$$\Gamma \subset 2\mu\,\Gamma_{1/2} \subset (2\mu)^2\,\{x \in \Gamma_{1/2} : u(x) < (\frac{1}{2} + \frac{1}{4})\min_\Gamma u\}.$$

Applying Lemma 5.4.3 to the set $\{x \in \Gamma_{1/2} : u(x) < (\frac{1}{2} + \frac{1}{4})\min_\Gamma u\}$ with $\beta = (\frac{1}{2} + \frac{1}{4})\min_\Gamma u$, we get

$$\Gamma \subset (2\mu)^3\,\{x \in \Gamma : u(x) < (\frac{1}{2} + \frac{1}{4} + \frac{1}{8})\min_\Gamma u\}.$$

Continuing in this way

$$\Gamma \subset (2\mu)^k\,\{x \in \Gamma : u(x) < (\frac{1}{2} + \frac{1}{4} + \frac{1}{8} + \cdots + \frac{1}{2^k})\min_\Gamma u\}$$

$$= (2\mu)^k\,\{x \in \Gamma : u(x) < (1 - \frac{1}{2^k})\min_\Gamma u\},$$

and the corollary is proved. ∎

Theorem 5.4.5 *If Γ is convex and u is a solution to $\lambda \leq Mu \leq \Lambda$ in Γ with $u = 0$ on $\partial\Gamma$, then u is $C^{1,\alpha}$ in the interior of Γ.*

Proof. We proceed in a sequence of steps.

Step 1. If Γ is normalized, then u is $C^{1,\alpha}$ at its minimum.

Let x_0 be the point where u attains its minimum. We shall prove that u is $C^{1,\alpha}$ at x_0. We have $u(x_0) = \min_\Gamma u \approx C(n, \lambda, \Lambda)$. Then by Aleksandrov's estimate, Theorem 1.4.2, $\mathrm{dist}(x_0, \partial\Gamma) > \epsilon_0$, and so $B_{\epsilon_0}(x_0) \subset \Gamma$. Let $x \in \Gamma$, $x \neq x_0$, and pick $k \geq 1$ such that

$$-u(x_0)\,2^{-k} \leq u(x) - u(x_0) < -u(x_0)\,2^{-k+1}.$$

So $u(x) \geq u(x_0) \left(1 - \dfrac{1}{2^k}\right)$, and consequently $x \notin \Gamma_{1/2^k}$. Then by Corollary 5.4.4, $x \notin (2\mu)^{-k} \Gamma$. So $x \notin (2\mu)^{-k} B_{\epsilon_0}(x_0) = B_{\epsilon_0/(2\mu)^k}(x_0)$, i.e., $|x - x_0| \geq \epsilon_0 (2\mu)^{-k}$. Since $\Gamma_{1/2} \subset \Gamma$ and by Lemma 5.4.1, $\frac{1}{2}\Gamma \subset \mu \Gamma_{1/2}$, it follows that $\mu > 1/2$. Then $\mu = 2^{-\theta}$ with $0 < \theta < 1$. So $|x - x_0| \geq \epsilon_0 (2 \cdot 2^{-\theta})^{-k} = \epsilon_0 (2^{-k})^{1-\theta}$. From $u(x) - u(x_0) < 2^{-(k-1)} (-u(x_0))$, we get $2^{-k} > \dfrac{u(x) - u(x_0)}{-2\, u(x_0)}$, and consequently $\epsilon_0 (2^{-k})^{1-\theta} \geq \epsilon_0 \left(\dfrac{u(x) - u(x_0)}{-2\, u(x_0)}\right)^{1-\theta}$. Therefore

$$0 \leq u(x) - u(x_0) \leq C(\epsilon_0, n, \lambda, \Lambda)\, |x - x_0|^{1/(1-\theta)}.$$

Step 2. If Γ is not necessarily normalized, then u is $C^{1,\alpha}$ at its minimum.

Again, let x_0 be the point where u attains its minimum. Normalize Γ with T, affine. Set $T(\Gamma) = \Gamma^*$ and $u^*(y) = |\det T|^{2/n} u(T^{-1}y)$. Then $\lambda \leq Mu^* \leq \Lambda$ and $u^* = 0$ on $\partial\Gamma^*$. We have $\min_{\Gamma^*} u^* = |\det T|^{2/n} \min_\Gamma u = |\det T|^{2/n} u(x_0) = u^*(Tx_0)$. We also have $\mathrm{dist}(Tx_0, \partial\Gamma^*) \geq C(n, \lambda, \Lambda)$ since Γ^* is normalized and the minimum is at Tx_0. Then

$$0 \leq u^*(y) - u^*(Tx_0) \leq C\, |y - Tx_0|^{1+\alpha}$$

by Step 1, and consequently

$$0 \leq u(x) - u(x_0) \leq \dfrac{C}{|\det T|^{2/n}}\, |Tx - Tx_0|^{1+\alpha}.$$

Letting $Tx = Ax + b$, we get

$$0 \leq u(x) - u(x_0) \leq \dfrac{C}{|\det T|^{2/n}}\, \|A\|^{1+\alpha}\, |x - x_0|^{1+\alpha}.$$

Step 3. If Γ is normalized, and $\mathrm{dist}(x_0, \partial\Gamma) \geq \epsilon_0$, then $|u(x) - \ell_{x_0}(x)| \leq C(\lambda, \Lambda, n, \epsilon_0)\, |x - x_0|^{1+\alpha}$, with ℓ_{x_0} a supporting hyperplane to u at x_0.

By Theorem 5.3.3, there exists $\delta = \delta(\epsilon_0, \lambda, \Lambda, n)$ such that $\Gamma_{x_0,\delta} = \{x \in \Gamma : u(x) < \ell_{x_0}(x) + \delta\} \Subset \Gamma$. Let $v(x) = u(x) - \ell_{x_0}(x) - \delta$. Then $\lambda \leq Mv \leq \Lambda$ in $\Gamma_{x_0,\delta}$, $v = 0$ on $\partial\Gamma_{x_0,\delta}$, and $v(x_0) = -\delta = \min_{\Gamma_{x_0,\delta}} v$. By Step 2,

$$0 \leq v(x) - v(x_0) \leq C(\epsilon_0, \lambda, \Lambda)\, \|A\|^{1+\alpha}\, |\det T|^{-2/n}\, |x - x_0|^{1+\alpha},$$

where $Tx = Ax + b$ normalizes $\Gamma_{x_0,\delta}$. To prove the claim we need to estimate $\|A\|$ and $|\det T|$. Let $u^*(y) = v(T^{-1}y)|\det T|^{2/n}$. We have $u^* = 0$ on $\partial\Gamma_{x_0,\delta}^*$ with $\Gamma_{x_0,\delta}^* = T(\Gamma_{x_0,\delta})$, and from (3.2.2) $\lambda \leq Mu^* \leq \Lambda$. Then by Proposition 3.2.3, we get $\min_{\Gamma_{x_0,\delta}^*} u^* \approx C(n, \lambda, \Lambda)$, but $\min_{\Gamma_{x_0,\delta}^*} u^* = -\delta\,|\det T|^{2/n}$ and so $|\det T| \approx \delta^{-n/2}$. It remains to estimate $\|A\|$. Let E be the ellipsoid of minimum volume containing $\Gamma_{x_0,\delta}$ and suppose E has axes of length $\lambda_1, \ldots, \lambda_n$. The matrix

A maps E to a translate of the unit ball. After a rotation we can assume A is diagonal, $A = diag\{\mu_1, \ldots, \mu_n\}$. Assume $b = 0$, then $A(E) = B_1(0)$, and so $(0, \ldots, 0, \lambda_i, 0, \ldots, 0) \in \partial E$ implies

$$A(0, \ldots, 0, \lambda_i, 0, \ldots, 0) = (0, \ldots, 0, \mu_i \lambda_i, 0, \ldots, 0) \in B_1(0).$$

That is $\mu_i = \lambda_i^{-1}$. Then $\|A\| = \max_i\{\lambda_i^{-1}\}$, and since $\det A = \lambda_1^{-1} \cdots \lambda_n^{-1} \approx \delta^{-n/2}$ and $\lambda_i \leq 1$, we get that $\|A\| \leq C\delta^{-n/2}$. Therefore

$$0 \leq v(x) - v(x_0) \leq C(\epsilon_0, \lambda, \Lambda, n)|x - x_0|^{1+\alpha},$$

for all $x \in \Gamma_{x_0,\delta}$.

Step 4. If Γ is not necessarily normalized, then the same conclusion as in Step 3 holds.

Normalize Γ with T affine and apply Step 3 to $u^*(y) = |\det T|^{2/n} u(T^{-1}y)$ in $T(\Gamma)$. Notice that the constant now will also depend on the eccentricity of Γ. ∎

Lemma 5.4.6 *Let Γ be strictly convex, not necessarily normalized, and $f \in C(\partial\Gamma)$. Let*

$$\gamma(f)(x) = \sup\{\ell(x) : \ell \text{ is affine and } \ell \leq f \text{ in } \partial\Gamma\}.$$

Suppose u is a convex solution to $\lambda \leq Mu \leq \Lambda$ in Γ with $u = f$ on $\partial\Gamma$. If $u(x_0) \lneqq \gamma(f)(x_0)$, then u is $C^{1,\alpha}$ at x_0.

Proof. We may assume that $u(x_0) \leq \gamma(f)(x_0) - \epsilon$ for some $\epsilon > 0$. We claim that if $\text{dist}(x_0, \partial\Gamma) \geq \epsilon_0$, then there exists $\delta > 0$ such that

$$\{x \in \Gamma : u(x) < \ell_{x_0}(x) + \delta\} \Subset \Gamma.$$

If we prove the claim, then the result follows from Steps 3 and 4 above. Suppose by contradiction that the claim is false. Then given $\delta = 1/k$, there exist $x_k \in \Gamma$ with $\text{dist}(x_k, \partial\Gamma) \geq \epsilon_0$, a solution u_k and a supporting hyperplane ℓ_{x_k} to u_k at x_k such that

$$\{x \in \Gamma : u_k(x) < \ell_{x_k}(x) + \frac{1}{k}\} \not\Subset \Gamma,$$

and $u_k(x_k) \leq \gamma(f)(x_k) - \epsilon$. Then there exists $y_k \in \partial\Gamma$ such that

$$f(y_k) = u_k(y_k) \leq \ell_{x_k}(y_k) + \frac{1}{k}. \tag{5.4.3}$$

We want a subsequence $u_k \to u_\infty$ uniformly on compact subsets of Γ such that $u_\infty = f$ on $\partial\Gamma$ and $u_\infty \in C(\bar{\Gamma})$. Since the u_k are convex, we have $u_k(x) \leq \max_\Gamma f$ for all k and $x \in \Gamma$. By Theorem 1.6.2, let v solve $Mv = \Lambda \geq Mu_k$ in Γ; $v = f$ on $\partial\Gamma$. By the comparison principle, Theorem 1.4.6, $u_k \geq v$ in Γ. Therefore the u_k are uniformly bounded in Γ. Hence by Lemma 3.2.1, $\partial u_k(\Gamma')$

is a uniformly bounded set in k for $\overline{\Gamma'} \subset \Gamma$. Therefore as in the proof of Lemma 5.3.1, we can select a subsequence of u_k such that $u_k \to u_\infty$ and we get that $\lambda \leq Mu_\infty \leq \Lambda$. We now select subsequences $x_k \to x_\infty$, dist$(x_\infty, \partial\Gamma) \geq \epsilon_0$, $y_k \to y_\infty \in \partial\Gamma$ and ℓ_{x_∞} a supporting hyperplane to u_∞ at x_∞. Passing to the limit in (5.4.3), we obtain $f(y_\infty) \leq \ell_{x_\infty}(y_\infty)$, and since ℓ_{x_∞} is a supporting hyperplane to u_∞, it follows that $\ell_{x_\infty}(y_\infty) = u_\infty(y_\infty)$. Also $\ell_{x_\infty}(x_\infty) = u_\infty(x_\infty)$, and therefore $u_\infty = \ell_{x_\infty}$ on the segment $\overline{x_\infty y_\infty}$. Hence $\overline{x_\infty y_\infty} \subset E = \{x \in \Gamma : u_\infty(x) = \ell_{x_\infty}(x)\}$, and by Theorem 5.2.1, $E^* \subset \partial\Gamma$, where E^* is the set of extremal points of E. Then $u_\infty(z) = \ell_{x_\infty}(z) = f(z)$ for $z \in E^*$. Since ℓ_{x_∞} is a supporting hyperplane to u_∞ and $u_\infty = f$ on $\partial\Gamma$, we have $\ell_{x_\infty}(x) \leq f(x)$ for $x \in \partial\Gamma$ and so $\ell_{x_\infty}(x) \leq \gamma(f)(x)$ for $x \in \Gamma$. At $z \in E^*$ we have $f(z) = \ell_{x_\infty}(z) \leq \gamma(f)(z)$, therefore $\ell_{x_\infty}(z) = \gamma(f)(z)$. Now write $x_\infty = \sum_{i=1}^N \lambda_i z_i$ with $z_i \in E^*$. Then

$$\gamma(f)(x_\infty) \leq \sum_{i=1}^N \lambda_i \gamma(f)(z_i) = \sum_{i=1}^N \lambda_i \ell_{x_\infty}(z_i) = \ell_{x_\infty}(x_\infty) = u_\infty(x_\infty).$$

On the other hand, passing to the limit in $u_k(x_k) \leq \gamma(f)(x_k) - \epsilon$ yields $u_\infty(x_\infty) \leq \gamma(f)(x_\infty) - \epsilon$, a contradiction. ∎

Theorem 5.4.7 *Let u be the convex solution to $\lambda \leq Mu \leq \Lambda$, $u = f$ on $\partial\Gamma$ with $f \in C^{1,\beta}(\partial\Gamma)$ and $\beta > 1 - \frac{2}{n}$. Then u is strictly convex in Γ.*

Proof. Suppose u is not strictly convex and let $x_0, x_1 \in \Gamma$ be such that the segment $\overline{P_0 P_1}$ is contained in the graph of u with $P_0 = (x_0, u(x_0))$ and $P_1 = (x_1, u(x_1))$. Let ℓ be a supporting hyperplane to u at the point $(x_0 + x_1)/2$, and let $E = \{x \in \Gamma : u(x) - \ell(x) = 0\}$. We have $\overline{x_0 x_1} \subset E$ and by Theorem 5.2.1, we have that $E^* \subset \partial\Gamma$, with E^* the set of extremal points of E. We claim that there are at least two distinct points $z_0, z_1 \in E^*$ such that $\overline{z_0 z_1} \cap E \neq \emptyset$. In fact, it is impossible that $\overline{z_0 z_1} \cap E = \emptyset$ for all pairs $z_0, z_1 \in E^*$ because by convexity $u - \ell = 0$ on $\overline{z_0 z_1}$, and since $x_0, x_1 \in E$, the segment $\overline{x_0 x_1}$ is generated by points in E^*. By rotating and translating the coordinates we may assume that $\overline{z_0 z_1}$ lies on the coordinate axis x_1 and $(z_0 + z_1)/2$ is the origin. Let $u^*(x) = u(x) - \ell(x)$. We have $u^* \geq 0$ and $\lambda \leq Mu^* \leq \Lambda$. We set $z_0 = -t_0 e_1$, $z_1 = t_0 e_1$ and we have $u^* = 0$ on $\overline{z_0 z_1}$. Let $x' = (x_2, \ldots, x_n)$. We shall construct T_ϵ a thin tube of diameter 2ϵ intersected with Γ around the segment $\overline{z_0 z_1}$, see Figure 5.2, and a barrier $B(t, x')$ such that $B(t, x') > u^*$ on ∂T_ϵ, $B(0,0) = 0$, and $MB < Mu^*$. Then by the comparison principle, Theorem 1.4.6, $B > u^*$ in T_ϵ, and consequently $u^*(0) < 0$, a contradiction.

Let

$$B(t, x') = C\left(a^{n-1} t^2 + \frac{1}{a} |x'|^2\right),$$

where C and a are constants to be determined. We first claim that

$$0 \leq u^*(t, x') \leq C_1 |x'|^{1+\beta}, \qquad \text{for } (t, x') \in T_\epsilon. \tag{5.4.4}$$

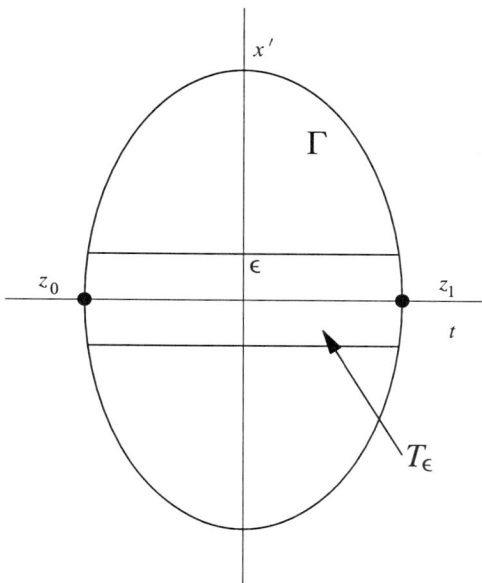

Figure 5.2. T_ϵ

Given $(t, x') \in T_\epsilon$, we write $(t, x') = \theta z'_0 + (1 - \theta)z'_1$ with $z'_0, z'_1 \in T_\epsilon \cap \partial\Gamma$, and so $0 \leq u^*(t, x') \leq \theta f^*(z'_0) + (1 - \theta)f^*(z'_1)$, where f^* is the corresponding data after changing the coordinates. Now $f^*(z_i) = u^*(z_i) = 0$ for $i = 0, 1$, and since $u^* \geq 0$ on $\partial\Gamma$ we have that $Df^*(z_0) = Df^*(z_1) = 0$. Since $f^* \in C^{1,\beta}(\partial\Gamma)$, we then have $f^*(z'_0) - f^*(z_0) = o(|z_0 - z'_0|^{1+\beta})$, and since $\partial\Gamma$ is Lipschitz, we have $|z'_0 - z_0| \leq c|x'|$ and consequently $f^*(z'_0) \leq c\,|x'|^{1+\beta}$. Analogously, $f^*(z'_1) \leq c\,|x'|^{1+\beta}$, and (5.4.4) is then proved.

Next we compare u^* and B on ∂T_ϵ. First on the lateral side of ∂T_ϵ. Here $|t| \leq t_0$ (roughly) and $|x'| = \epsilon$, and we then have $B(t, x') \geq C\,\epsilon^2 \dfrac{1}{a}$. On the other hand, by (5.4.4) we get $u^*(t, x') \leq C_1\,\epsilon^{1+\beta}$. Consequently to get $u^* < B$ on the lateral side we need

$$C\,\epsilon^2 \frac{1}{a} > C_1\,\epsilon^{1+\beta}. \tag{5.4.5}$$

On $T_\epsilon \cap \partial\Gamma$, we have $|t| \approx t_0$ and so $B(t, x') \geq C\,a^{n-1}\,t^2 \geq \dfrac{C}{2}\,a^{n-1}\,t_0^2$. Again by (5.4.4) we get $u^*(t, x') \leq C_1\,\epsilon^{1+\beta}$, and therefore we need that

$$\frac{C}{2}\,a^{n-1}\,t_0^2 > C_1\,\epsilon^{1+\beta}. \tag{5.4.6}$$

We now have that $MB = (2C)^n$ and if we pick C so that $(2C)^n < \lambda$ we obtain $MB < Mu^*$.

We choose $a = \dfrac{\epsilon^{1-\beta}}{\gamma}$ where γ is a large number to be determined in a moment. Inserting this value of a in (5.4.5) yields that we need $\gamma C > C_1$. Inserting the same value of a in (5.4.6) yields that

$$\frac{C}{2\gamma^{n-1}C_1} t_0^2 > \epsilon^{1+\beta+(\beta-1)(n-1)}. \tag{5.4.7}$$

The exponent of ϵ in (5.4.7) is positive if and only if $\beta > 1 - \frac{2}{n}$. Therefore picking ϵ sufficiently small in (5.4.7) yields the desired domain T_ϵ and the construction is complete. ∎

5.5 Examples

The following example shows that Theorem 5.4.7 is sharp. Let $x \in \mathbb{R}^n$ and set $x = (x_1, x')$; $r = \left(\sum_{i=2}^n x_i^2\right)^{1/2}$; $u(t, r)$ a function of two variables and $z(x) = z(x_1, \dots, x_n) = u(x_1, r)$. We have

$$\det D^2 z = \left(u_{tt} u_{rr} - (u_{tr})^2\right) \left(\frac{u_r}{r}\right)^{n-2}. \tag{5.5.1}$$

Assume that $n \geq 3$ and let $u(t, r) = (1 + t^2) r^\alpha$. We shall prove that for an appropriate choice of α, z is convex but not strictly convex and $\det D^2 z$ is bounded between two positive constants in a small ball. By (5.5.1),

$$\det D^2 z = 2\alpha^{n-1} (1+t^2)^{n-2} (\alpha - 1 - (\alpha + 1)t^2) r^{n(\alpha-2)+2}.$$

Pick α so that $n(\alpha - 2) + 2 = 0$, i.e., $\alpha = 2 - \dfrac{2}{n}$, then

$$\det D^2 z = 2 \left(2 - \frac{2}{n}\right)^{n-1} (1+x_1^2)^{n-2} \left(1 - \frac{2}{n} - \left(3 - \frac{2}{n}\right) x_1^2\right) \equiv \phi(x).$$

The function z is a generalized solution to the equation above that is strictly convex away from $x' = 0$. We show that $\lambda \leq Mz \leq \Lambda$ in a sufficiently small ball $B_\epsilon(0)$. Indeed, $\dfrac{1}{2}\left(1 - \dfrac{2}{n}\right) - \left(3 - \dfrac{2}{n}\right) x_1^2 > 0$ for $|x_1| < \sqrt{C_n} = \epsilon$. Notice that z is convex in $B_\epsilon(0)$ since it is continuous, nonnegative and convex away from $x' = 0$. By the choice of α we have that $z(x_1, \dots, x_n) = (1 + x_1^2) \left(\sum_{i=2}^n x_i^2\right)^{1-1/n}$, and so on the boundary of $B_\epsilon(0)$ we have $z(x) = (1 + x_1^2)(\epsilon^2 - x_1^2)^{1-1/n}$. We have that $z \in C^{1,1-2/n}(\partial B_\epsilon(0))$, but $z \notin C^{1,\beta}(\partial B_\epsilon(0))$ for $\beta > 1 - \frac{2}{n}$, showing that Theorem 5.4.7 is sharp.

In the same vein, we let $z(x_1, x') = |x'| + |x'|^\alpha (1 + x_1^2)$ with $\alpha = n/2$. As before, we have that Mz is bounded by two positive constants in a sufficiently small ball $B_\epsilon(0)$, z is convex but is not strictly convex, z is Lipschitz with constant 1, and Dz is only L^∞ and is not Hölder continuous.

5.6 Notes

The results in this chapter are due to Caffarelli, [Caf90b] and [Caf91]. The main result is the striking Theorem 5.2.1 about the extremal points of the set where u equals a supporting hyperplane. The examples in Section 5.5 are basically contained in [Pog78, pp. 81-86]. For counterexamples and results related to the $C^{1,\alpha}$ estimates of this chapter see [Caf93], [Wan92] and [Wan95].

Chapter 6

$W^{2,p}$ Estimates for the Monge–Ampère Equation

Our purpose in this chapter is to prove Caffarelli's interior L^p estimates for second derivatives of solutions to the Monge–Ampère equation. That is, solutions u to $Mu = f$ with f positive and continuous have second derivatives in L^p, for $0 < p < \infty$, Theorem 6.4.2. The origin of these estimates goes back to Pogorelov [Pog71] who proved that convex solutions to $\det D^2 u = 1$ on a bounded convex domain Ω with $u = 0$ on $\partial\Omega$ satisfy the L^∞ estimate

$$C_1(\Omega', \Omega)\, Id \le D^2 u(x) \le C_2(\Omega', \Omega)\, Id, \tag{6.0.1}$$

for $x \in \Omega'$, where Ω' is a convex domain with closure contained in Ω, Id is the identity matrix, and C_i are positive constants depending only on the domains. The estimates (6.0.1) have been proved in Chapter 4, and they follow as a consequence of Lemma 4.1.1; see (4.2.6).

6.1 Approximation Theorem

Theorem 6.1.1 *Assume Ω is a strictly convex domain with C^2 boundary such that $B_{1/n}(0) \subset \Omega \subset B_1(0)$, $0 < \epsilon < 1/2$, and u is a convex function in Ω that is a classical solution of*

$$1 - \epsilon \le \det D^2 u \le 1 + \epsilon, \qquad in\ \Omega \tag{6.1.1}$$
$$u = 0, \qquad on\ \partial\Omega.$$

Let $0 < \alpha < 1$ and define the set

$$\Omega_\alpha = \{x \in \Omega : u(x) < (1 - \alpha) \min_\Omega u\}.$$

Given a positive number λ define

$$A_\lambda = \{x_0 \in \Omega : \exists\ \ell(x)\ \textit{affine such that}$$

$$u(x) \geq \frac{\lambda}{2}|x - x_0|^2 + \ell(x), \forall x \in \Omega \textit{ and } \ell(x_0) = u(x_0)\}.$$

Then there exist a number $\sigma = \sigma(\alpha)$ and a constant $C = C_n$ depending only on the dimension n, and both independent of Ω, u and ϵ, such that

$$|\Omega_\alpha \setminus A_{\sigma(\alpha)}| < C_n\, \epsilon.$$

Proof. Let w be a solution of

$$\det D^2 w = 1 \qquad \text{in } \Omega,$$
$$w = 0 \qquad \text{on } \partial\Omega.$$

We have that $w \in C(\overline{\Omega}) \cap C^\infty(\Omega)$, see [CY77, Theorem 3, p. 59]. Also

$$\det D^2((1 + \epsilon)w) = (1 + \epsilon)^n \det D^2 w \geq (1 + \epsilon) \det D^2 w \geq \det D^2 u,$$

and by the comparison principle, Theorem 1.4.6,

$$(1 + \epsilon)w \leq u \qquad \text{in } \Omega. \tag{6.1.2}$$

Also,

$$\det D^2((1 - \epsilon)w) = (1 - \epsilon)^n \det D^2 w \leq (1 - \epsilon) \det D^2 w \leq \det D^2 u,$$

and consequently

$$(1 - \epsilon)w \geq u \qquad \text{in } \Omega.$$

So we get the estimate

$$(1 + \epsilon)w \leq u \leq (1 - \epsilon)w \qquad \text{in } \Omega.$$

Thus

$$\left(\frac{1}{2} + \epsilon\right)w \leq u - \frac{w}{2} \leq \left(\frac{1}{2} - \epsilon\right)w \qquad \text{in } \Omega. \tag{6.1.3}$$

Since $w < 0$ in Ω,

$$\left(\frac{1}{2} + \epsilon\right)(-w) \geq |u - \frac{w}{2}| \geq \left(\frac{1}{2} - \epsilon\right)(-w) \qquad \text{in } \Omega.$$

Since Ω is normalized and $w|_{\partial\Omega} = 0$, we have by Proposition 3.2.3 that $|\min_\Omega w| \approx 1$ and consequently

$$\max_\Omega |u - \frac{w}{2}| \approx 1.$$

Let $\Gamma(x)$ be the convex envelope of $u - \dfrac{w}{2}$ in Ω.

Claim 1. For all $x \in \Omega$ we have the inequality

$$|\frac{w(x)}{2} - \Gamma(x)| \leq C_n \epsilon.$$

In fact, by (6.1.3) and the fact that w is convex we have

$$\left(\frac{1}{2} + \epsilon\right) w(x) \leq \Gamma(x) \leq \left(\frac{1}{2} - \epsilon\right) w(x) \qquad \text{in } \Omega, \tag{6.1.4}$$

which yields

$$\epsilon w(x) \leq \Gamma(x) - \frac{w(x)}{2} \leq -\epsilon w(x),$$

i.e, $|\Gamma(x) - \dfrac{w(x)}{2}| \leq \epsilon(-w(x)) \leq C_n \epsilon$, since $|\max_\Omega(-w(x))| \approx 1$.

Claim 2.

$$\partial(\frac{1}{2} - \epsilon)w(\Omega) \subset \partial\Gamma(\Omega) \subset \partial(\frac{1}{2} + \epsilon)w(\Omega).$$

By (6.1.4) and since $w = 0$ on $\partial\Omega$, the claim follows from Lemma 1.4.1.

Claim 3. $\Gamma \in C^{1,1}(\Omega)$ and $\det D^2\Gamma = 0$ a.e. outside of the contact set

$$\mathcal{C} = \{x \in \Omega : \Gamma(x) = u(x) - \frac{w(x)}{2}\}.$$

This follows from Proposition 6.6.1.

If $x_0 \in \mathcal{C}$, then the function $u - \dfrac{w}{2} - \Gamma$ attains its minimum 0 at the point x_0 and hence $D^2(u - \dfrac{w}{2})(x_0) \geq D^2\Gamma(x_0) \geq 0$, for a.e. $x_0 \in \mathcal{C}$. Now for A, B symmetric and nonnegative matrices, we have the inequality

$$(\det(A + B))^{1/n} \geq (\det A)^{1/n} + (\det B)^{1/n}.$$

Hence

$$\left(\det D^2(u - \frac{w}{2})(x_0)\right)^{1/n} + \left(\det D^2(\frac{w}{2})(x_0)\right)^{1/n} \leq \left(\det D^2 u(x_0)\right)^{1/n},$$

which implies

$$\left(\det D^2\Gamma(x_0)\right)^{1/n} \leq \left(\det D^2 u(x_0)\right)^{1/n} - \left(\det D^2(\frac{w}{2})(x_0)\right)^{1/n}$$

$$\leq (1 + \epsilon)^{1/n} - \frac{1}{2} \leq 1 + \epsilon - \frac{1}{2} = \frac{1}{2} + \epsilon.$$

This inequality is valid for a.e. point in $\mathcal{C}$. Then by Claim 2,

$$|\partial(\frac{1}{2} - \epsilon)w(\Omega)| \leq |\partial\Gamma(\Omega)|,$$

and so by Claim 3,

$$\left(\frac{1}{2}-\epsilon\right)^n \int_\Omega \det D^2 w(x)\, dx \le \int_{\mathcal{C}} \det D^2 \Gamma(x)\, dx \le \left(\frac{1}{2}+\epsilon\right)^n |\mathcal{C}|.$$

This yields the estimate

$$|\mathcal{C}| \ge \left(\frac{\frac{1}{2}-\epsilon}{\frac{1}{2}+\epsilon}\right)^n |\Omega| > (1-4\,\epsilon)|\Omega|,$$

and therefore

$$|\Omega \setminus \mathcal{C}| < 4\,\epsilon\,|\Omega|. \tag{6.1.5}$$

On the set Ω_α, the function w is regular and by Pogorelov's estimate (4.2.6)

$$m_\alpha\, \mathrm{Id} \le D^2 w(x) \le M_\alpha\, \mathrm{Id} \qquad \forall x \in \Omega_\alpha, \tag{6.1.6}$$

with constants m_α, M_α depending only on α and n. Hence, if $x_0 \in \Omega_\alpha$, then by the convexity of w, we have the estimate

$$w(x) \ge w(x_0) + Dw(x_0) \cdot (x - x_0) + \frac{1}{2}m|x - x_0|^2,$$

for all $x \in \Omega$ and m a positive constant depending only on n and α. This inequality follows from the Taylor formula. Indeed,

$$w(x) - w(x_0) - Dw(x_0) \cdot (x - x_0)$$
$$= \int_0^1 (Dw(x_0 + t(x - x_0)) - Dw(x_0)) \cdot (x - x_0)\, dt,$$

and

$$(Dw(z) - Dw(y)) \cdot (z - y) = \int_0^1 \langle D^2 w(y + \theta(z - y))(z - y), z - y\rangle\, d\theta,$$

imply that

$$w(x) - w(x_0) - Dw(x_0) \cdot (x - x_0)$$
$$= \int_0^1 t \int_0^1 \langle D^2 w(x_0 + \theta\, t(x - x_0))(x - x_0), x - x_0\rangle\, d\theta dt$$
$$\ge \int_0^1 t \int_0^{1/2} \langle D^2 w(x_0 + \theta\, t(x - x_0))(x - x_0), x - x_0\rangle\, d\theta dt$$
$$\ge \int_0^{1/2} m\,|x - x_0|^2\, d\theta,$$

by (6.1.6), with $m = m(\alpha)$; because from the convexity of w and (6.1.2) we have that $x_0 + \theta t (x - x_0) \in \Omega_{\alpha'}$, with $\alpha' = \frac{2}{3} + \frac{1}{3}\alpha$, for all $0 \le \theta \le 1/2, 0 \le t \le 1$, and $x \in \Omega$.

On the other hand, since Γ is the convex envelope of $u(x) - \dfrac{w(x)}{2}$, we have $u(x) - \dfrac{w(x)}{2} \ge \Gamma(x), \forall x \in \Omega$. Since Γ is convex and $x_0 \in C$, let ℓ_{x_0} be a supporting hyperplane to Γ at x_0. Then

$$u(x) \ge \ell_{x_0}(x) + \frac{w(x)}{2} \ge \ell(x) + \frac{1}{4}m|x - x_0|^2, \qquad \forall x \in \Omega.$$

This means that given $0 < \alpha < 1$, there exists a constant σ, depending only on α and n, such that

$$\Omega_\alpha \cap C \subset A_\sigma \cap \Omega_\alpha, \qquad \sigma = \frac{m}{2}, \tag{6.1.7}$$

which implies

$$\Omega_\alpha \setminus A_\sigma \subset \Omega_\alpha \setminus C.$$

Therefore from (6.1.5) we get

$$|\Omega_\alpha \setminus A_{\sigma(\alpha)}| < C_n \epsilon, \tag{6.1.8}$$

and the theorem follows.   ∎

6.2   Tangent paraboloids

The following lemma shows that if a solution to Monge–Ampère has a touching paraboloid from below, then it has locally a touching paraboloid from above.

Lemma 6.2.1 *Let u be convex in Ω such that $\lambda \le Mu \le \Lambda$, $u \ge 0$ in Ω, $u(x_0) = 0$ and*
$$u(x) \ge \sigma_0|x - x_0|^2, \qquad \forall x \in \Omega.$$
Then there exists a constant $C = C(n, \lambda, \Lambda)$ such that
$$u(x) \le C \sigma_0^{-n+1}|x - x_0|^2,$$
for $|x - x_0| \le \delta$ with δ sufficiently small.

Proof. Let $S_t = \{x \in \Omega : u(x) < t\}$. Noticing that S_t is a section, by Theorem 5.3.3 and Corollary 3.2.4, we get that $|S_t| \approx t^{n/2}$, for $t \le \delta$. Now

$$\{x \in \Omega : u(x) < t\} \subset \{x \in \Omega : \sigma_0|x - x_0|^2 < t\} \subset B_{\sqrt{t/\sigma_0}}(x_0).$$

Let $v(x) = u(x) - t$. We have $v = 0$ on ∂S_t, and by the Aleksandrov maximum principle,

$$|v(x)|^n \le C_n \operatorname{dist}(x, \partial S_t) \, (\operatorname{diam}(S_t))^{n-1} \, Mv(S_t).$$

Taking $x = x_0$ yields

$$t^n \le C(n, \lambda, \Lambda) \operatorname{dist}(x_0, \partial S_t) \left(\sqrt{t/\sigma_0} \right)^{n-1} t^{n/2}.$$

Therefore $\operatorname{dist}(x_0, \partial S_t) \ge C(n, \lambda, \Lambda) \sigma_0^{\frac{n-1}{2}} \sqrt{t}$, and so

$$B_{c(n,\lambda,\Lambda)\sqrt{t\sigma_0^{n-1}}}(x_0) \subset S_t,$$

for $t \le \delta$. Hence, if $x \in \Omega$, then $x \notin S_{u(x)}$ and if $u(x) \le \delta$ we have $x \notin B_{c_n\sqrt{u(x)\sigma_0^{n-1}}}(x_0)$. That is $|x - x_0| \ge c_n \sqrt{u(x)\sigma_0^{n-1}}$ and the lemma follows. ∎

We assume that $\lambda \le Mu \le \Lambda$ in Ω. If $0 < \alpha \le \alpha_0 < 1$ then there exists $\eta_0 > 0$ depending only on α_0 such that $S_t(x_0) \subset \Omega_{(\alpha_0+1)/2}$ for all $t \le \eta_0$ and $x_0 \in \Omega_\alpha$. Then given $\lambda > 0$ we define

$$D_\lambda^\alpha = \{x_0 \in \overline{\Omega_\alpha} : S_t(x_0) \subset B_{\lambda\sqrt{t}}(x_0) \text{ for all } t \le \eta_0\}.$$

Also

$$A_\sigma(u) = \{x_0 \in \Omega : u(x) \ge \sigma|x - x_0|^2 + Du(x_0) \cdot (x - x_0) + u(x_0), \quad \forall x \in \Omega\}.$$

Lemma 6.2.2 *There exists a constant $C_1 > 0$ depending only on α_0, λ, Λ, n, and* $\operatorname{diam} \Omega$ *such that*

$$D_\lambda^\alpha = \overline{\Omega_\alpha} \cap A_{1/\lambda^2}(u),$$

for all $\lambda \ge C_1$ and $0 < \alpha \le \alpha_0 < 1$.

Proof.   By Theorem 5.3.3, there exists an $\eta_0 > 0$ depending only on the dimension (note that $\Omega_{1/2}$ and the set of points x such that $\operatorname{dist}(x, \partial\Omega) \ge 1/4$ are comparable) such that

$$S_t(x_0) \Subset \Omega, \qquad \forall x_0 \in \Omega_{1/2}, \qquad \forall t \le \eta_0.$$

Let $x_0 \in D_\lambda^\alpha$ and ℓ_{x_0} be a supporting hyperplane to u at x_0. Given $z \in \Omega$, let $\mu = u(z) - \ell_{x_0}(z) (\ge 0)$. Then $z \in \overline{S_\mu(x_0)}$. If $\mu \le \eta_0$, then $S_\mu(x_0) \subset B_{\lambda\sqrt{\mu}}(x_0)$, that is, $|z - x_0|^2 \le \lambda^2(u(z) - \ell_{x_0}(z))$, which implies

$$u(z) \ge \frac{1}{\lambda^2}|z - x_0|^2 + \ell_{x_0}(z).$$

If, on the other hand, $\mu > \eta_0$, then

$$u(z) - \ell_{x_0}(z) > \eta_0 = \frac{\eta_0}{\operatorname{diam}(\Omega)^2}\operatorname{diam}(\Omega)^2 \ge \frac{\eta_0}{\operatorname{diam}(\Omega)^2}|z - x_0|^2,$$

for all $z \in \Omega$. So, if $\dfrac{\eta_0}{\text{diam}(\Omega)^2} \geq \dfrac{1}{\lambda^2}$, then

$$u(z) \geq \frac{1}{\lambda^2}|z - x_0|^2 + \ell_{x_0}(z), \qquad \forall z \in \Omega,$$

that is, $x_0 \in A_{1/\lambda^2}(u)$.

If $x_0 \in \overline{\Omega_\alpha} \cap A_{1/\lambda^2}(u)$ and $x \in S_t(x_0)$ with $t \leq \eta_0$, then $u(x) - \ell_{x_0}(x) \leq t$ and therefore $t \geq \dfrac{1}{\lambda^2}|x - x_0|^2$. This completes the proof of the lemma. ∎

6.3 Density estimates and power decay

Proposition 6.3.1 *Let $0 < \epsilon < 1/2$, and u a solution of $1 - \epsilon \leq Mu \leq 1 + \epsilon$ in the normalized convex domain Ω with $u = 0$ on $\partial\Omega$. Then there exists a constant $c_0 > 0$ depending only on n and σ in the Approximation Theorem 6.1.1 such that if $x_0 \in \Omega_{\alpha_0}$ and $h \leq \eta_0/2$, then we have*

$$\frac{|S_h(x_0) \setminus A_{c_0 h}(u)|}{|S_h(x_0)|} < C_n \epsilon.$$

Moreover, if $\lambda \geq \dfrac{2}{c_0 \eta_0}$, then

$$\frac{|S_h(x_0) \setminus A_{1/\lambda}(u)|}{|S_h(x_0)|} < C_n \epsilon,$$

for $h \geq \dfrac{1}{c_0 \lambda}$; ($C_n$ is the constant in the approximation Theorem 6.1.1).

Proof. The idea of the proof is to normalize u and then apply the approximation Theorem 6.1.1 on a section. Notice that by Remark 3.3.9 $S_h(x_0)$ is strictly convex, and since u is a classical solution, the boundary $\partial S_h(x_0)$ is C^2. Let T be the affine transformation that normalizes $S_h(x_0)$. That is,

$$B_{\alpha_n}(0) \subset T\left(S_h(x_0)\right) \subset B_1(0).$$

We have $S_h(x_0) = \{x : u(x) < \ell_{x_0}(x) + h\}$ with $\ell_{x_0}(x) = u(x_0) + Du(x_0) \cdot (x - x_0)$. Let

$$v(x) = \frac{C}{h}(u(x) - \ell_{x_0}(x) - h),$$

where C is a constant that will be determined in a moment. Set

$$u^*(y) = v(T^{-1}y),$$

and

$$S_h^*(x_0) = T\left(S_h(x_0)\right).$$

We have

$$D^2 u^*(y) = \frac{C}{h} (T^{-1})^t (D^2 u)(T^{-1}(y))(T^{-1}).$$

Hence

$$\det D^2 u^*(y) = \left(\frac{C}{h}\right)^n |\det T^{-1}|^2 \det D^2 u(T^{-1}(y)).$$

We now pick C such that

$$\frac{C^n}{h^n} |\det T^{-1}|^2 = 1.$$

Since u satisfies the equation $1 - \epsilon \le Mu \le 1 + \epsilon$, it follows that u^* satisfies

$$1 - \epsilon \le \det D^2 u^* \le 1 + \epsilon \qquad \text{in } S_h^*(x_0),$$
$$u^* = 0 \qquad \text{on } \partial S_h^*(x_0).$$

By definition of u^* we have that

$$\min_{S_h^*(x_0)} u^* = -C.$$

By Corollary 3.2.4, we have that $|S_h(x_0)| \approx h^{n/2}$, for $h < \delta$ depending only on the structure and, since $|T(S_h(x_0))| \approx 1$, it follows that $|\det T| |S_h| \approx 1$ and consequently $|\det T| \approx h^{-n/2}$. Therefore $C \approx C_n$. We then apply the Approximation Theorem 6.1.1 with $\Omega \to S_h^*(x_0) = S^*$, $\alpha \to \beta$, and $u \to u^*$ and we obtain

$$\frac{|S_{\beta h}^*(x_0) \setminus A_\sigma(u^*)|}{|S_{\beta h}^*(x_0)|} < C_n \epsilon, \qquad \text{with } T(S_{\beta h}(x_0)) = S_{\beta h}^*(x_0).$$

Notice that $(S^*)_\beta = (S_h^*(x_0))_\beta = T(S_{\beta h}(x_0))$, and the doubling property implies that $|S_{\beta h}^*(x_0)| \approx C(\beta)$. We now show that there exist universal constants $0 < \beta < 1$ and $c_0 > 0$ such that

$$S_{\beta h}^*(x_0) \cap A_\sigma(u^*) \subset T\left(S_{\beta h}(x_0) \cap A_{c_0 h}(u)\right). \tag{6.3.1}$$

Let $x^* \in S_{\beta h}^*(x_0) \cap A_\sigma(u^*)$. Then $x^* = T x_1$ with $x_1 \in S_{\beta h}(x_0)$. Since $x^* \in A_\sigma(u^*)$, we have that

$$u^*(x) \ge u^*(x^*) + Du^*(x^*) \cdot (x - x^*) + \sigma |x - x^*|^2,$$

for all $x \in S_h^*(x_0)$. Changing variables back we obtain

$$u(x) - \ell_{x_1}(x) \ge \sigma \frac{h}{C} |Tx - Tx_1|^2,$$

for all $x \in S_h(x_0)$. Recall that if $x_h = $ center of mass of $S_h(x_0)$ and E is the ellipsoid of minimum volume centered at x_h and containing $S_h(x_0)$, the transformation T is such that $T(S_h(x_0)) \subset T(E) = B_1(0)$. Rotating the coordinates,

we may assume that the ellipsoid E has axes on the coordinate axes. That is,
$$Tx = (\frac{x_1 - x_h^1}{\mu_1}, \ldots, \frac{x_n - x_h^n}{\mu_n})$$ where μ_i are the axes of the ellipsoid. Since
$E \subset 3\Omega$ and Ω is bounded, we have that $\mu_i \leq const$, and so $\mu_i^{-1} \geq const$.
Therefore $|Tx - Tx_1| \geq C'|x - x_1|$. Consequently,

$$u(x) - \ell_{x_1}(x) \geq C'\sigma h |x - x_1|^2, \qquad \text{in } S_h(x_0). \tag{6.3.2}$$

We now want to show that a similar inequality holds in all Ω. Since $x_1 \in S_{\beta h}(x_0)$, by the engulfing property, Theorem 3.3.7, $S_{\beta h}(x_0) \subset S_{\theta \beta h}(x_1)$ and hence $x_0 \in S_{\theta \beta h}(x_1)$. Again by the engulfing property $S_{\theta \beta h}(x_1) \subset S_{\theta^2 \beta h}(x_0)$. If we pick $\beta = 1/\theta^2$, then $S_{h/\theta}(x_1) \subset S_h(x_0)$, and the inequality (6.3.2) holds for $x \in S_{h/\theta}(x_1)$. If $x \notin S_{h/\theta}(x_1)$, then $u(x) - \ell_{x_1}(x) \geq h/\theta$, and we write

$$\frac{h}{\theta} = h \frac{\sigma}{\theta} \frac{1}{\sigma} \frac{|x - x_1|^2}{|x - x_1|^2} \geq C(\theta, \sigma) h \sigma |x - x_1|^2,$$

since $|x - x_1| \leq \text{diam}(\Omega)$. That is $x_1 \in A_{\bar{C}\sigma h}(u)$ and therefore (6.3.1) follows with $c_0 = \bar{C}\sigma$, and $\beta = 1/\theta^2$.

By (6.3.1),

$$(S^*)_\beta \setminus A_\sigma(u^*) = (S^*)_\beta \setminus ((S^*)_\beta \cap A_\sigma(u^*)) \supset (S^*)_\beta \setminus T(S_{\beta h}(x_0) \cap A_{c_0 h}(u)).$$

Applying T^{-1} to both terms and setting $L = S_{\beta h}(x_0) \cap A_{c_0 h}(u)$ yields

$$T^{-1}((S^*)_\beta \setminus A_\sigma(u^*)) \supset T^{-1}((S^*)_\beta \setminus T(L)).$$

Since $T^{-1}(A \setminus B) = T^{-1}(A) \setminus T^{-1}(B)$, we get

$$T^{-1}((S^*)_\beta \setminus A_\sigma(u^*)) \supset T^{-1}((S^*)_\beta) \setminus L$$
$$= S_{\beta h}(x_0) \setminus (S_{\beta h}(x_0) \cap A_{c_0 h}(u)) = S_{\beta h}(x_0) \setminus A_{c_0 h}(u).$$

Consequently

$$\frac{|S_{\beta h}(x_0) \setminus A_{c_0 h}(u)|}{|S_{\beta h}(x_0)|} \leq \frac{|T^{-1}(S_{\beta h}^*(x_0) \setminus A_\sigma(u^*))|}{|T^{-1}(S_{\beta h}^*(x_0))|} = \frac{|S_{\beta h}^*(x_0) \setminus A_\sigma(u^*)|}{|S_{\beta h}^*(x_0)|} < C_n \epsilon,$$

for $h < \eta_0$, which yields the first conclusion of the proposition.

To prove the second conclusion, notice that if $\sigma \geq \mu$, then $A_\sigma(u) \subset A_\mu(u)$. Hence $A_{c_0 h}(u) \subset A_{1/\lambda}(u)$ for $1/\lambda \leq c_0 h$. Now $h \leq \eta_0/2$ and if we assume $h \geq 1/c_0\lambda$ we have $\eta_0/2 \geq 1/c_0\lambda$, which forces us to choose $\lambda \geq 2/c_0\eta_0$. $\blacksquare$

Theorem 6.3.2 *Let $0 < \epsilon < 1/2$; $1 - \epsilon \leq Mu \leq 1 + \epsilon$ in Ω normalized with $u|_{\partial\Omega} = 0$.. There exist positive constants M, p_0, and C_2, independent of ϵ, such that*
$$|\Omega_\tau \setminus D_{M\lambda}^\tau| \leq \sqrt{C_n \epsilon} |\Omega_\alpha \setminus D_\lambda^\alpha|,$$
for all $\lambda \geq C_2$ and $0 < \tau < \alpha \leq \alpha_0$ with $\alpha - \tau = (M\lambda)^{-p_0}$.

Proof. Let $\mathcal{O} = \Omega_\tau \setminus D^\tau_{M\lambda}$ and notice that $\mathcal{O}$ is open since $D^\tau_{M\lambda}$ is closed. By Lemma 6.2.2, $D^\tau_{M\lambda} = \overline{\Omega_\tau} \cap A_{1/(M\lambda)^2}(u)$ for each λ such that $M\lambda \geq C_1$ and $0 < \tau \leq \alpha_0$. Hence $\mathcal{O} = \Omega_\tau \setminus A_{1/(M\lambda)^2}(u)$, and so $S_h(x_0) \cap \mathcal{O} \subset S_h(x_0) \setminus A_{1/(M\lambda)^2}(u)$. Then by Proposition 6.3.1,

$$\frac{|S_h(x_0) \cap \mathcal{O}|}{|S_h(x_0)|} < C_n \epsilon, \tag{6.3.3}$$

for $x_0 \in \Omega_{\alpha_0}$, $M\lambda \geq \max\left\{C_1, \sqrt{\frac{2}{C_0\eta_0}}\right\}$, and $\frac{1}{(M\lambda)^2} \leq h \leq \frac{\eta_0}{2}$. On the other hand, since $\mathcal{O}$ is open, we have that

$$\lim_{h \to 0} \frac{|S_h(x_0) \cap \mathcal{O}|}{|S_h(x_0)|} = 1, \qquad \text{for } x_0 \in \mathcal{O}.$$

We now use the following theorem.

Theorem 6.3.3 (Covering Theorem) *Let $\mathcal{O}$ be a bounded open set and $0 < \epsilon$ small. Suppose that for each $x \in \mathcal{O}$ a section $S_{h_x}(x)$ is given with $h_x \leq \beta$, and*

$$\frac{|S_{h_x}(x) \cap \mathcal{O}|}{|S_{h_x}(x)|} = C_n \epsilon.$$

Then there exists a subfamily of sections $\{S_{h_k}(x_k)\}_{k=1}^\infty$ such that

(1) $x_k \in \mathcal{O}$ and $h_k \leq \beta$ for all k;

(2) $\mathcal{O} \subset \cup_{k=1}^\infty S_{h_k}(x_k)$;

(3) $\dfrac{|S_{h_k}(x_k) \cap \mathcal{O}|}{|S_{h_k}(x_k)|} = C_n \epsilon$; and

(4) $|\mathcal{O}| \leq \sqrt{C_n \epsilon} \, |\cup_{k=1}^\infty S_{h_k}(x_k)|$.

We postpone the proof of this theorem to Section 6.5.

If $\mathcal{O} = \Omega_\tau \setminus D^\tau_{M\lambda}$, then for $x \in \mathcal{O}$ we choose h_x to be the largest h such that $\dfrac{|S_h(x) \cap \mathcal{O}|}{|S_h(x)|} \geq C_n \epsilon$. From (6.3.3), $h_x \leq \dfrac{1}{(M\lambda)^2}$. By Lemma 6.5.1 the ratio $|S_h(x) \cap \mathcal{O}|/|S_h(x)|$ is a continuous function of h, and consequently $|S_{h_x}(x) \cap \mathcal{O}|/|S_{h_x}(x)| = C_n \epsilon$. Applying the covering theorem to $\mathcal{O}$ and $\beta = \dfrac{1}{(M\lambda)^2}$ we obtain a family of sections $S_{h_k}(x_k)$ satisfying (1)–(4).

Claim: There exist M, p_0, positive constants, such that

$$S_{h_k}(x_k) \subset \Omega_\alpha \setminus D^\alpha_\lambda, \qquad \text{for all } k, \tag{6.3.4}$$

for $0 < \tau < \alpha \leq \alpha_0$ with $\alpha - \tau = (M\lambda)^{-p_0}$.

We first show that

$$S_{h_k}(x_k) \subset \Omega_\alpha, \qquad \text{for all } k, \tag{6.3.5}$$

for $0 < \tau < \alpha \leq \alpha_0$ with $\alpha - \tau = (M\lambda)^{-p_0}$. We apply Theorem 3.3.10
(i) with $t = -u(x_0) = -\min_\Omega u$, $r = \tau$, $s = \alpha$, and noticing that $\Omega_\alpha = S_u(x_0, \alpha(-u(x_0)))$ and $\Omega_\tau = S_u(x_0, \tau(-u(x_0)))$. Since $x_k \in \Omega_\tau$, $S_u(x_k, C_0(\alpha - \tau)^{p_1} u(x_0)) \subset S_u(x_0, -\alpha u(x_0))$. Since Ω is normalized, we have that $u(x_0) \approx C_n$.
If $C_0(\alpha - \tau)^{p_1} u(x_0) \approx \dfrac{1}{(M\lambda)^2}$, then we obtain (6.3.5).

Second we prove that

$$S_{h_k}(x_k) \subset (D_\lambda^\alpha)^c. \tag{6.3.6}$$

Suppose by contradiction that there exists $x_0 \in S_{h_k}(x_k) \cap D_\lambda^\alpha$. By the engulfing
property, $S_{h_k}(x_k) \subset S_{\theta h_k}(x_0)$. Since $h_k \leq \dfrac{1}{(M\lambda)^2}$, we can pick λ sufficiently
large such that $\theta 2 h_k \leq \eta_0$, and since $x_0 \in D_\lambda^\alpha$ we get

$$S_{h_k}(x_k) \subset S_{2h_k}(x_k) \subset S_{\theta 2 h_k}(x_0) \subset B(x_0, \lambda\sqrt{2\theta h_k}) \subset B(x_k^*, 2\lambda\sqrt{2\theta h_k}),$$

with $x_k^* = $ center of mass of $S_{2h_k}(x_k)$. Let T be an affine transformation normalizing the section $S_{2h_k}(x_k)$,

$$B_{\alpha_n}(0) \subset T(S_{2h_k}(x_k)) = S_k^* \subset B_1(0),$$

with $Tx_k^* = 0$. Then $B_{\alpha_n}(0) \subset T\left(B(x_k^*, 2\lambda\sqrt{2\theta h_k})\right)$. We claim that

$$T^{-1}(B(y_0, h)) \subset B\left(T^{-1} y_0, 4\alpha_n^{-1}\lambda h\sqrt{\theta h_k}\right), \tag{6.3.7}$$

for any $h > 0$ and y_0. Indeed, given $z \in B(y_0, h)$ we write $z = y_0 + e h$ with
$|e| < 1$. We have $e = \alpha_n^{-1}(\alpha_n e) = \alpha_n^{-1} T\beta_k$ with $\beta_k \in B(x_k^*, 2\lambda\sqrt{2\theta h_k})$. Now
$Tx = A(x - x_k^*)$ and $T^{-1}x = A^{-1}x + x_k^*$. Then

$$|T^{-1}z - T^{-1} y_0| = |A^{-1}(e h)| = \alpha_n^{-1} h |A^{-1}(e\alpha_n)|$$
$$= \alpha_n^{-1} h |A^{-1}(A(\beta_k - x_k^*))|$$
$$= \alpha_n^{-1} h |\beta_k - x_k^*| \leq \alpha_n^{-1} h 2\lambda\sqrt{2\theta h_k},$$

and (6.3.7) is proved.

Let $v(x) = \dfrac{C}{2h_k}\left(u(x) - \ell_{x_k}(x) - 2h_k\right)$, and $u^*(y) = v(T^{-1} y)$ with C as in
the previous proposition. We have that $1 - \epsilon \leq Mu^* \leq 1 + \epsilon$ and $u^* = 0$ on ∂S_k^*.
By the Approximation Theorem 6.1.1

$$|(S_k^*)_{1/2} \setminus A_\sigma(u^*)| < C_n \epsilon |(S_k^*)_{1/2}|. \tag{6.3.8}$$

Notice that $T(S_{h_k}(x_k)) = (S_k^*)_{1/2}$. We claim that

$$T^{-1}\left((S_k^*)_{1/2} \cap A_\sigma(u^*)\right) \subset D_{M\lambda}^\alpha, \tag{6.3.9}$$

for M sufficiently large. Let $y^* \in A_\sigma(u^*)$ with $y^* = Ty$, $y \in S_{h_k}(x_k)$. We have

$$u^*(x^*) - \ell_{y^*}^*(x^*) \geq \sigma |x^* - y^*|^2, \qquad \text{for each } x^* \in S_k^*.$$

So if $x^* \in S_{u^*}(y^*, r)$ and $r \leq \delta$, where δ is the number in Theorem 5.3.3, then $|x^* - y^*| \leq \sqrt{r/\sigma}$, i.e.,

$$S_{u^*}(y^*, r) \subset B(y^*, \sqrt{r/\sigma}), \qquad \text{for } r \leq \delta' \text{ with } \delta' \leq \delta. \tag{6.3.10}$$

By definition of v, $T^{-1}(S_{u^*}(y^*, r)) = S_u(y, 2rh_k/C)$. Then by (6.3.7)

$$T^{-1}(B(y^*, \sqrt{r/\sigma})) \subset B(y, 4\alpha_n^{-1}\lambda\sqrt{\theta h_k}\sqrt{r/\sigma}),$$

and consequently

$$S_u(y, 2rh_k/C) \subset B(y, 4\alpha_n^{-1}\lambda\sqrt{\theta h_k}\sqrt{r/\sigma}). \tag{6.3.11}$$

We shall prove that $y \in D_{M\lambda}^\alpha$ for M large, i.e., $y \in \Omega_\alpha$ and $S_u(y, h) \subset B(y, M\lambda\sqrt{h})$. From (6.3.5), $y \in \Omega_\alpha$.

Case 1. Assume that $\dfrac{C h}{2h_k} \leq \delta' \leq \delta$. Applying (6.3.11) with $r = \dfrac{C h}{2 h_k}$ yields

$$S_u(y, h) \subset B(y, 4\alpha_n^{-1}\lambda\sqrt{C\theta/2\sigma}\sqrt{h}),$$

and the claim follows with $M = 4\alpha_n^{-1}\sqrt{C\theta/2\sigma}$.

Case 2. Assume that $\dfrac{C h}{2h_k} \geq \delta'$, and $h \leq \eta_0$. We pick $\delta' = \min\{\delta, \theta C/2\}$. Hence $h_k \leq \dfrac{C}{2\delta'}h$ and $y \in S_u\left(x_k, \dfrac{C}{2\delta'}h\right)$. By the engulfing property, Theorem 3.3.7 and since $x_0 \in S_{h_k}(x_k)$, it follows that

$$y, x_0 \in S_u\left(x_k, \frac{C}{2\delta'}h\right) \subset S_u\left(y, \frac{\theta C}{2\delta'}h\right).$$

Again by the engulfing property,

$$S_u\left(y, \frac{\theta C}{2\delta'}h\right) \subset S_u\left(x_0, \frac{\theta^2 C}{2\delta'}h\right).$$

If $\dfrac{\theta^2 C}{2\delta'}h \leq \eta_0$, then

$$S_u\left(x_0, \frac{\theta^2 C}{2\delta'}h\right) \subset B\left(x_0, \lambda\sqrt{\frac{\theta^2 C}{2\delta'}h}\right)$$

since $x_0 \in D_\lambda^\alpha$. So

$$S_u\left(y, \frac{\theta C}{2\delta'} h\right) \subset B\left(y, 2\lambda \sqrt{\frac{\theta^2 C}{2\delta'}} h\right).$$

By the choice of δ' we have $\dfrac{\theta C}{2\delta'} \geq 1$ and consequently

$$S_u(y, h) \subset B\left(y, 2\lambda \sqrt{\frac{\theta^2 C}{2\delta'}} h\right) \quad \text{for } h \leq \frac{2\delta' \eta_0}{\theta^2 C} = \eta_0' < \eta_0.$$

If $\eta_0' < h \leq \eta_0$, then

$$S_u(y, h) \subset \Omega \subset B(y, \operatorname{diam}(\Omega)) = B\left(y, \frac{\operatorname{diam}(\Omega)}{\sqrt{h}} \sqrt{h}\right)$$

$$\subset B\left(y, \frac{\operatorname{diam}(\Omega)}{\sqrt{\eta_0'}} \sqrt{h}\right).$$

Therefore $y \in D_{M\lambda}^\alpha$ for

$$M \geq \max\left\{2\sqrt{\frac{\theta^2 C}{2\delta'}}, \frac{\operatorname{diam}(\Omega)}{\sqrt{\eta_0'}}\right\},$$

and $\lambda \geq 1$. This completes the proof of (6.3.9).

Finally, with the aid of (6.3.9) we shall contradict (3) in the Covering Theorem 6.3.3. We write

$$\begin{aligned}
S_{h_k}(x_k) \cap \mathcal{O} &= S_{h_k}(x_k) \cap (\Omega_\tau \setminus D_{M\lambda}^\tau) \\
&= S_{h_k}(x_k) \cap \Omega_\tau \cap (D_{M\lambda}^\tau)^c \\
&= S_{h_k}(x_k) \cap \Omega_\tau \cap \{\overline{\Omega_\tau} \cap A_{1/(M\lambda)^2}(u)\}^c \\
&= S_{h_k}(x_k) \cap \Omega_\tau \cap \{(\overline{\Omega_\tau})^c \cup (A_{1/(M\lambda)^2}(u))^c\} \\
&= S_{h_k}(x_k) \cap \Omega_\tau \cap (A_{1/(M\lambda)^2}(u))^c \\
&\subset S_{h_k}(x_k) \cap \Omega_\alpha \cap (A_{1/(M\lambda)^2}(u))^c = (*).
\end{aligned}$$

It follows from Lemma 6.2.2 that $\Omega_\alpha \cap (A_{1/(M\lambda)^2}(u))^c \subset (D_{M\lambda}^\alpha)^c$. Therefore

$$(*) \subset S_{h_k}(x_k) \cap (D_{M\lambda}^\alpha)^c = (**).$$

We claim that $(**) \subset T^{-1}\left((S_k^*)_{1/2} \setminus A_\sigma(u^*)\right)$. In fact, by (6.3.9)

$$(S_k^*)_{1/2} \setminus A_\sigma(u^*) = (S_k^*)_{1/2} \setminus \left((S_k^*)_{1/2} \cap A_\sigma(u^*)\right) \supset (S_k^*)_{1/2} \setminus T\left(D_{M\lambda}^\alpha\right).$$

Applying T^{-1} yields

$$T^{-1}\left((S_k^*)_{1/2} \setminus A_\sigma(u^*)\right) \supset T^{-1}\left((S_k^*)_{1/2} \setminus T\left(D_{M\lambda}^\alpha\right)\right)$$
$$= T^{-1}\left((S_k^*)_{1/2}\right) \setminus D_{M\lambda}^\alpha = S_{h_k}(x_k) \setminus D_{M\lambda}^\alpha,$$

and the claim is proved.

Therefore

$$\frac{|S_{h_k}(x_k) \cap \mathcal{O}|}{|S_{h_k}(x_k)|} \leq \frac{|T^{-1}\left((S_k^*)_{1/2} \setminus A_\sigma(u^*)\right)|}{|T^{-1}\left((S_k^*)_{1/2}\right)|}$$
$$= \frac{|(S_k^*)_{1/2} \setminus A_\sigma(u^*)|}{|(S_k^*)_{1/2}|} < C_n \epsilon,$$

by (6.3.8). This contradicts (3) in the Covering Theorem 6.3.3 and completes the proof of the theorem. ∎

6.4 L^p estimates of second derivatives

We first prove the $W^{2,p}$ estimates with data zero and right-hand side sufficiently close to 1.

Theorem 6.4.1 *Let Ω be a normalized convex domain. Let $0 < \epsilon < 1/2$ and suppose u is a convex classical solution to (6.1.1) with $u = 0$ on $\partial\Omega$. Then given $0 < p < \infty$ and $0 < \tau < \alpha \leq \alpha_0$, there exists $\epsilon(p, \tau) > 0$ such that*

$$\int_{\Omega_\tau} D_{ee}u(x)^p \, dx \leq C,$$

for all $|e| = 1$ and $0 < \epsilon < \epsilon(p, \tau)$, with C a constant depending only on $p, \tau,$ and n.

Proof.   We iterate the inequality in Theorem 6.3.2. Notice that we can pick M large so that the statement of that theorem holds for all $\lambda \geq M$. We then begin the iteration with $\lambda = M$ and let $\alpha_1 = \alpha - M^{-2p_0}$. We get

$$|\Omega_{\alpha_1} \setminus D_{M^2}^{\alpha_1}| \leq \sqrt{C_n \epsilon}\, |\Omega_\alpha \setminus D_M^\alpha|.$$

If $\lambda = M^2$ and $\alpha_2 = \alpha_1 - M^{-3p_0}$, then

$$|\Omega_{\alpha_2} \setminus D_{M^3}^{\alpha_2}| \leq \sqrt{\epsilon}\, |\Omega_{\alpha_1} \setminus D_{M^2}^{\alpha_1}| \leq (\sqrt{\epsilon})^2 |\Omega_\alpha \setminus D_M^\alpha|.$$

Continuing in this way we let $\alpha_k = \alpha_{k-1} - M^{-(k+1)p_0} = \alpha - \sum_{j=1}^k M^{-(j+1)p_0}$ and obtain

$$|\Omega_{\alpha_k} \setminus D_{M^{k+1}}^{\alpha_k}| \leq (\sqrt{C_n \epsilon})^k |\Omega_\alpha \setminus D_M^\alpha|.$$

We fix $\tau < \alpha$ and choose M even larger than before (now depending also on τ) so that $\alpha_k \geq \alpha - \sum_{j=1}^{\infty} M^{-(j+1)p_0} \geq \tau$. If $x_0 \in A_\gamma(u)$ then $u(x) \geq \gamma |x - x_0|^2 + \ell_{x_0}(x)$ for all $x \in \Omega$. Then by Lemma 6.2.1, $u(x) \leq C(n) \gamma^{-n+1} |x - x_0|^2 + \ell_{x_0}(x)$ for x close to x_0. That is $D_{ee}u(x_0) \leq 2 C(n) \gamma^{-(n-1)}$ where e is any unit vector. By Lemma 6.2.2, if $x_0 \in D_{M^{i+1}}^{\alpha_i}$, then $x_0 \in \overline{\Omega}_{\alpha_i} \cap A_{1/M^{2(i+1)}}(u)$ and consequently $D_{ee}u(x_0) \leq 2 C(n) M^{2(n-1)(i+1)}$. Therefore

$$D_{M^{i+1}}^{\alpha_i} \subset \{x_0 \in \overline{\Omega}_{\alpha_i} : D_{ee}u(x_0) \leq 2 C(n) M^{2(n-1)(i+1)}\},$$

and consequently

$$\{x_0 \in \overline{\Omega}_{\alpha_i} : D_{ee}u(x_0) > 2 C(n) M^{2(n-1)(i+1)}\} \subset \overline{\Omega}_{\alpha_i} \setminus D_{M^{i+1}}^{\alpha_i}.$$

Now we can estimate the L^p-norm of $D_{ee}u$. We have

$$\|D_{ee}u\|_{L^p(\Omega_\tau)}^p$$

$$\leq M^{2(n-1)p} |\Omega_\tau| + \sum_{i=0}^{\infty} \int_{\{x \in \Omega_\tau : M^{2(n-1)(i+1)} < D_{ee}u(x) \leq M^{2(n-1)(i+2)}\}} D_{ee}u(x)^p \, dx$$

$$\leq M^{2(n-1)p} |\Omega_\tau| + \sum_{i=0}^{\infty} \int_{\{x \in \overline{\Omega}_{\alpha_i} : M^{2(n-1)(i+1)} < D_{ee}u(x) \leq M^{2(n-1)(i+2)}\}} D_{ee}u(x)^p \, dx$$

$$\leq C(M, n, \alpha, \tau, p) + \sum_{i=0}^{\infty} |\overline{\Omega}_{\alpha_i} \setminus D_{M^{i+1}}^{\alpha_i}| \, M^{2(n-1)(i+2)p}$$

$$\leq C(M, n, \alpha, \tau, p) + C(n) \sum_{i=0}^{\infty} (\sqrt{C_n}\,\epsilon)^{i+1} \, M^{2(n-1)(i+2)p} < \infty,$$

for ϵ sufficiently small.   ∎

Notice that if A is an $n \times n$ symmetric matrix, then the eigenvalues of A, $\lambda_1 \geq \lambda_2 \geq \cdots \geq \lambda_n$, are given by the formula:

$$\lambda_{n-k+1} = \inf_{S_k} \max_{\{x \in S_k : |x| = 1\}} A x \cdot x, \qquad k = 1, \ldots, n,$$

where the infimum is taken over all subspaces S_k of $\mathbb{R}^n$ with dimension k; see [CH53, vol. I, p. 32]. Noticing that in the previous argument the vector e can be replaced by an arbitrary measurable function $e(x) : \Omega_\tau \to S^{n-1}$, we obtain that

$$\|\lambda_j(x)\|_{L^p(\Omega_\tau)} \leq C,$$

for $j = 1, \ldots, n$, where $\lambda_j(x)$ is the j-th eigenvalue of the matrix $D^2 u(x)$. This generalizes the estimate (4.2.6).

Theorem 6.4.2 *Let Ω be a normalized convex domain and $f \in C(\bar{\Omega})$ with $0 < \lambda \leq f(x) \leq \Lambda$ in Ω. Suppose u is a solution to $Mu = f$ in Ω with $u|_\Omega = 0$. Then for each $0 < p < \infty$ and $0 < \alpha < 1$ we have*

$$\int_{\Omega_\alpha} D_{ee}u(x)^p \, dx \leq C(n, p, \alpha),$$

for all $|e| = 1$.

Proof.   Let $y \in \Omega$ and suppose that we have a section $S = S_u(y, \delta) \subset \Omega$ such that $|f(y) - f(x)| \leq \epsilon$, for each $x \in S_u(y, \delta)$. Notice that since Ω is normalized we have from the property of size of sections, Theorem 3.3.8, that

$$B(y, K_1 \delta) \subset S(y, \delta) \subset B(y, K_2 \delta^b), \tag{6.4.1}$$

with K_1, K_2, b positive constants depending only on λ, Λ and n. Let T be an affine transformation normalizing $S_\delta(y)$ and consider the function

$$v(x) = \frac{|\det T|^{2/n}}{f(y)^{1/n}} \left(u(T^{-1}x) - \ell_y(T^{-1}x) - \delta \right),$$

where ℓ_y is the supporting hyperplane to u at y. We look at v on the set $T(S_u(y, \delta))$, and we have $v = 0$ on $\partial T(S_u(y, \delta))$, and

$$D^2 v(x) = \frac{|\det T|^{2/n}}{f(y)^{1/n}} \left\{ (T^{-1})^t \, (D^2 u)(T^{-1}x) \, T^{-1} \right\}.$$

Hence

$$Mv(x) = \frac{|\det T|^2}{f(y)} \, |\det T|^{-2} \, (Mu)(T^{-1}x) = \frac{f(T^{-1}x)}{f(y)}.$$

Now $f(y) - \epsilon \leq f(x) \leq f(y) + \epsilon$ for $x \in S$, and so

$$1 - \frac{\epsilon}{f(y)} \leq \frac{f(T^{-1}x)}{f(y)} \leq 1 + \frac{\epsilon}{f(y)},$$

for $x \in T(S)$. Since $f(y) \geq \lambda$, it follows that

$$1 - \frac{\epsilon}{\lambda} \leq \frac{f(T^{-1}x)}{f(y)} \leq 1 + \frac{\epsilon}{\lambda}, \qquad \text{for } x \in T(S).$$

Then applying Theorem 6.4.1 on the set $T(S)$ to the function v we get that

$$\int_{(T(S))_\alpha} D_{ee}v(x)^p \, dx \leq C(n, \alpha, p),$$

for each unit vector e and $\epsilon \leq \epsilon_{p,\alpha}$.

By definition of v we have that

$$D^2 u(x) = \frac{f(y)^{1/n}}{|\det T|^{2/n}} T^t (D^2 v)(Tx) T,$$

and consequently

$$\begin{aligned}
D_{ee}u(x) &= \langle D^2 u(x) e, e \rangle \\
&= \frac{f(y)^{1/n}}{|\det T|^{2/n}} \langle T^t (D^2 v)(Tx) Te, e \rangle \\
&= \frac{f(y)^{1/n}}{|\det T|^{2/n}} \langle (D^2 v)(Tx) Te, Te \rangle \\
&= \frac{f(y)^{1/n}}{|\det T|^{2/n}} |Te|^2 \langle (D^2 v)(Tx) e', e' \rangle \qquad e' = \frac{Te}{|Te|} \\
&= \frac{f(y)^{1/n}}{|\det T|^{2/n}} |Te|^2 (D_{e'e'}v)(Tx).
\end{aligned}$$

We have $S = \{x : u(x) - \ell_y(x) - \delta < 0\}$ and so $S_\alpha = S_u(y, \alpha \delta)$. On the other hand, since $T(S) = S_{u \circ T^{-1}}(Ty, \delta)$, it follows that $(T(S))_\alpha = T(S_\alpha)$. Also $S_{u \circ T^{-1}}(Ty, \alpha \delta) = \{x : v(x) < (1 - \alpha) \min_{T(S)} v\}$ by definition of v. Therefore

$$\begin{aligned}
\int_{S_\alpha} D_{ee}u(x)^p \, dx &= \frac{f(y)^{p/n}}{|\det T|^{2p/n}} |Te|^{2p} \int_{S_\alpha} (D_{e'e'}v)(Tx)^p \, dx \\
&= \frac{f(y)^{p/n}}{|\det T|^{2p/n}} |Te|^{2p} \int_{(T(S))_\alpha} (D_{e'e'}v)(z)^p |\det T|^{-1} \, dz \\
&\leq f(y)^{p/n} \left(\frac{|Te|^2}{|\det T|^{\frac{2}{n} + \frac{1}{p}}} \right)^p C(\alpha, n, p).
\end{aligned}$$

To estimate the term between parentheses, let E be the ellipsoid of minimum volume containing S, and let $\mu_1, \dots, \mu_n$ be the axes of E. If δ is small, then by Corollary 3.2.4 we have that $|S| \approx \delta^{n/2}$. The affine transformation that normalizes S has the form

$$Tx = \left(\frac{x_1 - x_1^0}{\mu_1}, \dots, \frac{x_n - x_n^0}{\mu_n} \right),$$

where $(x_1^0, \dots, x_n^0)$ is the center of the ellipsoid E (the center of mass of S). We have $|\det T| \approx \delta^{-n/2}$, and from (6.4.1) it follows that $\mu_i \geq K_1 \delta$. Hence

$$\frac{|Te|^2}{|\det T|^{\frac{2}{n} + \frac{1}{p}}} \approx |Te|^2 \delta^{1 + \frac{n}{2p}} \leq C \delta^{\frac{n}{2p} - 1},$$

and consequently

$$\int_{S_\alpha} D_{ee}u(x)^p \, dx \leq C(\lambda, \Lambda, n, \alpha, p) \delta^{\frac{n}{2} - p}. \tag{6.4.2}$$

We now pick δ small depending only on the parameters λ, Λ, α and the modulus of continuity of f, so that $|f(y) - f(x)| \leq \epsilon$ in $B(y, K_2 \delta^b)$, $y \in \Omega_\alpha$, and next select a finite covering of Ω_α by balls $\{B(y_j, K_1 \delta)\}_{j=1}^N$ with $y_j \in \Omega_\alpha$. The desired inequality then follows by adding (6.4.2) over $(S(y_j, \delta))_\alpha$.   ∎

6.5   Proof of the Covering Theorem 6.3.3

We begin by stating and proving two lemmas. The following lemma shows that the Lebesgue measure is doubling on the sections of a convex function.

Lemma 6.5.1 *Let ϕ be a convex function in $\mathbb{R}^n$ whose sections $S_\phi(x, p, t)$ are bounded sets. Then*

(a) $|S_\phi(x, p, t)| \leq 2^n |S_\phi(x, p, t/2)|.$

(b) *For all $0 < \delta < 1$, $|S_\phi(x, p, t)| - |S_\phi(x, p, \delta t)| \leq n(1 - \delta)|S_\phi(x, p, t)|.$*

Proof. Let $\ell(z)$ be a supporting hyperplane to ϕ at the point $(x, \phi(x))$ with slope p. We have $S_\phi(x, p, t) = \{z : \phi(z) \leq \ell(z) + t\}$. Let Γ be the cone in $\mathbb{R}^{n+1}$ with vertex $(x, \phi(x))$ and base B spanned by the set $\{(z, \phi(z)) : z \in \partial S_\phi(x, p, t)\}$. Since ϕ is convex, $\Gamma \subset \{(z, s) : \phi(z) \leq s\}$. The orthogonal projection of B on $\mathbb{R}^n$ is the section $S_\phi(x, p, t)$. Let $0 < \delta < 1$, $H_\delta = \Gamma \cap \{(z, \ell(z) + \delta t) : z \in \mathbb{R}^n\}$, and B_δ the orthogonal projection of H_δ on $\mathbb{R}^n$. The convexity of ϕ implies that $B_\delta \subset S_\phi(x, p, \delta t)$. Let Γ_δ be the cone with vertex $(x, \phi(x))$ and base H_δ. We have $|\Gamma_\delta| = \delta^{n+1} |\Gamma|$, $|\Gamma| = \dfrac{1}{n+1} h |B|$, and $|\Gamma_\delta| = \dfrac{1}{n+1} \delta h |H_\delta|$, where h is the height of the cone Γ. Hence $|H_\delta| = \delta^n |B|$. On the other hand, $|S_\phi(x, p, t)| = c_1 |B|$ and $|B_\delta| = c_1 |H_\delta|$. Therefore,

$$|S_\phi(x, p, t)| = \delta^{-n} |B_\delta| \leq \delta^{-n} |S_\phi(x, p, \delta t)|,$$

which is the doubling condition if $\delta = 1/2$. We also have

$$
\begin{aligned}
|S_\phi(x, p, t)| - |S_\phi(x, p, \delta t)| &\leq |S_\phi(x, p, t)| - |B_\delta| \\
&= |S_\phi(x, p, t)| - \delta^n |S_\phi(x, p, t)| \\
&\leq (1 - \delta^n) |S_\phi(x, p, t)| \leq n (1 - \delta) |S_\phi(x, p, t)|.
\end{aligned}
$$

∎

It is well known that the Besicovitch covering lemma (see [Ste93, §8.17, p.44]) does not hold for general families of convex sets. However, for the convex sets $S_u(x, t)$ we have the following result.

Lemma 6.5.2 *Suppose that u is a convex function whose cross-sections satisfy the conclusions of Theorem 3.3.8 and Corollary 3.3.6 (ii). Let $A \subset \mathbb{R}^n$ be a* *bounded set. Suppose that for each $x \in A$ a section $S_u(x, t)$ is given such that t is bounded by a fixed number M. Let us denote by $\mathcal{F}$ the family of all these sections. Then there exists a countable subfamily of $\mathcal{F}$, $\{S_u(x_k, t_k)\}_{k=1}^{\infty}$, with the following properties:*

(i) *$A \subset \cup_{k=1}^{\infty} S_u(x_k, t_k)$.*

(ii) *$x_k \notin \cup_{j<k} S_u(x_j, t_j)$, $\forall k \geq 2$.*

(iii) *For $0 < \epsilon < \epsilon_0$, with ϵ_0 sufficiently small depending only on the constants in (3.3.6) and (3.3.5), we have that the family*

$$\mathcal{F}_\epsilon = \{S_u(x_k, (1-\epsilon)t_k)\}_{k=1}^{\infty}$$

has bounded overlaps. More precisely,

$$\sum_{k=1}^{\infty} \chi_{S_u(x_k,(1-\epsilon)t_k)}(x) \leq K \, \log \frac{1}{\epsilon},$$

where K is a constant depending only on the constants in (3.3.6) and (3.3.5); χ_E denotes the characteristic function of the set E.

Proof. To simplify the notation we write $S_u(x, t) = S(x, t)$. We may assume $M = \sup\{t : S(x, t) \in \mathcal{F}\}$. Let

$$\mathcal{F}_0 = \{S(x, t) : \frac{M}{2} < t \leq M, \quad S(x, t) \in \mathcal{F}\},$$

and

$$A_0 = \{x : S(x, t) \in \mathcal{F}_0\}.$$

Pick $S(x_1, t_1) \in \mathcal{F}_0$ such that $t_1 > \frac{3}{4}M$. Then either $A_0 \setminus S(x_1, t_1) = \emptyset$ or $A_0 \setminus S(x_1, t_1) \neq \emptyset$. In the first case, $A_0 \subset S(x_1, t_1)$ and we stop. In the second case, the set

$$\{t : S(x, t) \in \mathcal{F}_0 \text{ and } x \in A_0 \setminus S(x_1, t_1)\}$$

is nonempty and we let α_2 denote its supremum. Pick t_2 in this set such that $\alpha_2 \geq t_2 > \frac{3}{4}\alpha_2$ and let $S(x_2, t_2)$ be the corresponding section. We then have $x_2 \notin S(x_1, t_1)$ and $t_1 > \frac{3}{4}M \geq \frac{3}{4}\alpha_2 > \frac{3}{4}t_2$. Again, we have either $A_0 \setminus (S(x_1, t_1) \cup S(x_2, t_2)) = \emptyset$ or $A_0 \setminus (S(x_1, t_1) \cup S(x_2, t_2)) \neq \emptyset$. In the first case, we have $A_0 \subset S(x_1, t_1) \cup S(x_2, t_2)$ and we stop. In the second case, we continue the process. In general, for the jth-stage we pick t_j such that $\alpha_j \geq t_j > \frac{3}{4}\alpha_j$ where

$$\alpha_j = \sup\{t : S(x, t) \in \mathcal{F}_0 \text{ and } x \in A_0 \setminus \bigcup_{i<j} S(x_i, t_i)\},$$

and select $S(x_j, t_j)$. We have $t_i > (\frac{3}{4})^{j-i}t_j$ for $j > i$. Continuing in this way we construct a family, possibly infinite, which we denote by

$$\mathcal{F}_0' = \{S(x_k^0, t_k^0)\}_{k=1}^{\infty}.$$

with

$$x_j^0 \in A_0 \setminus \bigcup_{i<j} S(x_i^0, t_i^0).$$

We now consider the family $\mathcal{F}_1 = \{S(x,t) : \frac{M}{4} < t \leq \frac{M}{2}\}$. Let

$$A_1 = \{x : S(x,t) \in \mathcal{F}_1 \quad \text{and} \quad x \notin \bigcup_{i=1}^{\infty} S(x_i^0, t_i^0)\}.$$

We repeat the construction above for the set A_1, obtaining a family of sections denoted by

$$\mathcal{F}_1' = \{S(x_i^1, t_i^1)\}_{i=1}^{\infty}.$$

We continue this process and in the kth-stage we consider the family $\mathcal{F}_k = \{S(x,t) : \frac{M}{2^{k+1}} < t \leq \frac{M}{2^k}\}$ and the set

$$A_k = \{x : S(x,t) \in \mathcal{F}_k \quad \text{and} \quad x \notin \bigcup \text{sections previously selected}\}.$$

In the same way as before, we obtain a family of sections denoted by

$$\mathcal{F}_k' = \{S(x_i^k, t_i^k)\}_{i=1}^{\infty}.$$

Obviously, each section $S(x_i^k, t_i^k)$ in the generation $\mathcal{F}_k'$ has the property that

$$\frac{M}{2^{k+1}} < t_i^k \leq \frac{M}{2^k}.$$

We claim that the collection of all the sections in all generations $\mathcal{F}_k'$, $k \geq 1$, is the family that satisfies the conclusions of the lemma.

To show (i), we shall first prove that each generation $\mathcal{F}_k'$ has overlapping bounded by a constant depending only on the parameters in (3.3.6), and independent of k and M. Second, we shall deduce from this that each generation $\mathcal{F}_k'$ has a finite number of members; in particular, by relabeling the members of $\mathcal{F}_k'$ we obtain (ii). This implies that the process in the construction of $\mathcal{F}_k'$ stopped at some point and therefore all the points of A_k are covered by the union of $\mathcal{F}_k'$. Consequently, (i) follows.

Let us then show that each generation $\mathcal{F}_k'$ has bounded overlapping. Suppose that

$$z_0 \in S(x_{j_1}^k, t_{j_1}^k) \cap \cdots \cap S(x_{j_N}^k, t_{j_N}^k),$$

with $S(x_{j_i}^k, t_{j_i}^k) \in \mathcal{F}_k'$. To simplify the notation we set $x_{j_i}^k = x_i$, $t_{j_i} = t_i$, and let t_0 be the maximum of all these t_i, $1 \leq i \leq N$. We may assume by construction that $x_l \notin S(x_i, t_i)$ for $l > i$. By (3.3.6) we have that

$$B(z_i, K_2 t_i / t_0) \subset T(S(x_i, t_i)) \subset B\left(z_i, K_1 \left(\frac{t_i}{t_0}\right)^{\epsilon_1}\right),$$

for $1 \le i \le N$, where T is an affine transformation that normalizes $S(x_0, t_0)$, $z_i = T(x_i)$, and $|z_i| \le K_3$. Since $\frac{M}{2^{k+1}} < t_i \le \frac{M}{2^k}$, we get

$$B(z_i, K_2/2) \subset T(S(x_i, t_i)) \subset B(z_i, K_1).$$

Since $x_l \notin S(x_i, t_i)$, we have that

$$T(x_l) \notin T(S(x_i, t_i)), \qquad l > i,$$

and consequently,

$$|T(x_l) - T(x_i)| > K_2/2, \qquad l > i. \tag{6.5.1}$$

Let Q be the cube in $\mathbb{R}^n$ with center 0 and edgelength $2(K_1 + K_3)$. We have $T(S(x_i, t_i)) \subset Q$, $1 \le i \le N$. We divide the cube Q into α^n congruent subcubes $\tilde{Q}$ with edgelength $2(K_1 + K_3)/\alpha$. If $T(x_k), T(x_m) \in \tilde{Q}$ then

$$|T(x_k) - T(x_m)| < 2\sqrt{n}\,\frac{K_1 + K_3}{\alpha}.$$

If we select α large such that $2\sqrt{n}\,\dfrac{K_1 + K_3}{\alpha} < K_2/2$, then by (6.5.1) each subcube $\tilde{Q}$ contains at most one $T(x_i)$. Therefore the overlapping in each generation $\mathcal{F}'_k$ is at the most α^n, where α can be taken to be the smallest integer bigger than $4\sqrt{n}\,\dfrac{K_1 + K_3}{K_2}$.

Let us now prove that the family $\mathcal{F}'_k = \{S(x^k_i, t^k_i)\}_{i=1}^{\infty}$ is finite. We set again for simplicity $x^k_i = x_i$ and $t^k_i = t_i$. Since A is bounded, let $C \ge M/2^k$ be a constant such that $A \subset S(z, C)$ for some z. From (3.3.6) we obtain

$$B(z_i, K_2 t_i/C) \subset T(S(x_i, t_i)) \subset B\left(z_i, K_1 \left(\frac{t_i}{C}\right)^{\epsilon_1}\right),$$

where T is an affine transformation that normalizes $S(z, C)$, and $z_i = Tx_i$ with $|z_i| \le K_3$. We have that

$$\frac{M}{2^k C} < \frac{t_i}{C} \le 1,$$

and therefore

$$B\left(z_i, K_2\frac{M}{2^k C}\right) \subset T(S(x_i, t_i)) \subset B(0, K_1 + K_3).$$

Since the family $\mathcal{F}'_k$ has overlapping bounded by α^n, the family $T(S(x_i, t_i))$ also has overlapping bounded by α^n. Then

$$\sum_i \chi_{T(S(x_i, t_i))}(x) \le \alpha^n,$$

which implies

$$\sum_i \chi_{B(z_i, K_2 M/2^k C)}(x) \leq \alpha^n.$$

Integration of this inequality over the ball $B(0, K)$, where $K = K_1 + K_3$, yields

$$\sum_i \omega_n \left(\frac{K_2 M}{2^k C}\right)^n \leq \alpha^n \, \omega_n \, K^n,$$

which implies that the number of terms in the sum is finite and we are done.

We now estimate the overlapping of sections belonging to different generations, but for this we need to "shrink" the sections selected. Let $0 < \epsilon < 1$ and

$$z_0 \in \bigcap_i S(x_{r_i}^{e_i}, (1 - \epsilon)t_{r_i}^{e_i}), \tag{6.5.2}$$

where $e_1 < e_2 < \cdots < e_i < \cdots$, $M2^{-(e_i+1)} < t_{r_i}^{e_i} \leq M2^{-e_i}$, and for simplicity in the notation we set $x_i = x_{r_i}^{e_i}$ and $t_i = t_{r_i}^{e_i}$. Fix i and let $j > i$; we shall measure the gap between e_j and e_i. Let T_i be an affine transformation that normalizes the section $S(x_i, t_i)$. We have $t_i > t_j$ for $j > i$. Then again by (3.3.6)

$$B\left(z_{ij}, K_2 t_j/t_i\right) \subset T_i(S(x_j, t_j)) \subset B\left(z_{ij}, K_1 \left(\frac{t_j}{t_i}\right)^{\epsilon_1}\right),$$

where $z_{ij} = T_i(x_j)$ and $|z_{ij}| \leq K_3$. By construction $x_j \notin S(x_i, t_i)$. Then by (3.3.5) we have that

$$B(T_i(x_j), C \epsilon^n) \cap T_i(S(x_i, (1 - \epsilon)t_i)) = \emptyset.$$

Consequently,

$$C \epsilon^n < |T_i(x_j) - T_i(z_0)| \leq K_1 \left(\frac{t_j}{t_i}\right)^{\epsilon_1} \leq 2^{\epsilon_1} K_1 2^{(e_i - e_j)\epsilon_1}.$$

Hence

$$e_j - e_i \leq C_1 \ln \frac{1}{\epsilon},$$

where C_1 is a constant depending only on ϵ_1, n, C and K_1, for all $\epsilon > 0$ small, smallness depending only on the previous constants. In particular, the number of members in (6.5.2) is at most $C_1 \ln(1/\epsilon)$, which together with the fact previously proved that the overlapping of members in the same generation is at the most α^n gives (iii), and the proof of the lemma is complete. ∎

We can now prove Theorem 6.3.3.

Proof. (of Theorem 6.3.3) Set $\delta = C_n \epsilon$. Let us take $\mu > 0$ small and apply the variant of Besicovitch's covering lemma, Lemma 6.5.2, to the family of sections

$\mathcal{F} = \{S(x, h_x)\}_{x \in \mathcal{O}}$. Then we can select from $\mathcal{F}$ a countable subfamily, denoted by $\{S_k = S(x_k, h_k)\}_{k=1}^{\infty}$, such that $\mathcal{O} \subset \cup_{k=1}^{\infty} S(x_k, h_k)$ and

$$\sum_{k=1}^{\infty} \chi_{S(x_k, (1-\mu)h_k)}(x) \leq C_1 \, \log(1/\mu).$$

Let $n_N^{\mu}(x)$ be the overlapping function for the finite family $S_k^{\mu} = S(x_k, (1-\mu)h_k)$ with $k = 1, \ldots, N$, i.e.,

$$n_N^{\mu}(x) = \begin{cases} \#\{k : x \in S(x_k, (1-\mu)h_k)\}, & \text{if } x \in \cup_{k=1}^{N} S(x_k, (1-\mu)h_k), \\ 1, & \text{if } x \notin \cup_{k=1}^{N} S(x_k, (1-\mu)h_k). \end{cases}$$

We have the formula

$$\chi_{\cup_{k=1}^{N} S_k^{\mu}}(x) = \frac{1}{n_N^{\mu}(x)} \sum_{1}^{N} \chi_{S_k^{\mu}}(x). \tag{6.5.3}$$

By Lemma 6.5.2(iii) we have $n_N^{\mu}(x) \leq K \, \log(1/\mu)$.

We have

$$|\mathcal{O}| = |\mathcal{O} \cap \cup_{k=1}^{\infty} S_k| = \lim_{N \to \infty} |\mathcal{O} \cap \cup_{k=1}^{N} S_k|,$$

and since $|\mathcal{O} \cap S_k| = \delta \, |S_k|$, it follows that

$$|\mathcal{O} \cap \cup_{k=1}^{N} S_k| \leq \sum_{k=1}^{N} |\mathcal{O} \cap S_k| \leq \delta \sum_{k=1}^{N} |S_k|$$

$$= \delta \sum_{k=1}^{N} |S_k \setminus S_k^{\mu}| + \delta \sum_{k=1}^{N} |S_k^{\mu}|$$

$$= I + II.$$

We first estimate I. By Lemma 6.5.1(b), we have $|S_k \setminus S_k^\mu| \le n\,\mu\,|S_k|$ and consequently $\dfrac{|S_k|}{|S_k^\mu|} \le \dfrac{1}{1 - n\mu}$. Thus

$$I \le \delta\,n\,\mu \sum_{k=1}^{N} |S_k| \le \delta\,\frac{n\,\mu}{1 - n\mu} \sum_{k=1}^{N} |S_k^\mu|$$

$$= \delta\,\frac{n\,\mu}{1 - n\mu} \int \sum_{k=1}^{N} \chi_{S_k^\mu}(x)\,dx$$

$$= \delta\,\frac{n\,\mu}{1 - n\mu} \int n_N^\mu(x)\,\frac{1}{n_N^\mu(x)} \sum_{k=1}^{N} \chi_{S_k^\mu}(x)\,dx$$

$$\le \delta\,\frac{n\,\mu}{1 - n\mu}\,K\,\log(1/\mu) \int \frac{1}{n_N^\mu(x)} \sum_{k=1}^{N} \chi_{S_k^\mu}(x)\,dx$$

$$= \delta\,\frac{n\,\mu}{1 - n\mu}\,K\,\log(1/\mu)\,|\cup_{k=1}^{N} S_k^\mu|.$$

To estimate II we write

$$II = \delta \int \sum_{k=1}^{N} \chi_{S_k^\mu}(x)\,dx$$

$$= \delta \int n_N^\mu(x)\,\frac{1}{n_N^\mu(x)} \sum_{k=1}^{N} \chi_{S_k^\mu}(x)\,dx$$

$$\le \delta\,K\,\log(1/\mu) \int \frac{1}{n_N^\mu(x)} \sum_{k=1}^{N} \chi_{S_k^\mu}(x)\,dx$$

$$= \delta\,K\,\log(1/\mu)\,|\cup_{k=1}^{N} S_k^\mu|.$$

Therefore

$$I + II \le \delta\,K\,\log(1/\mu) \left(1 + \frac{n\,\mu}{1 - n\mu}\right) |\cup_{k=1}^{N} S_k^\mu|$$

$$\le 2\,\delta\,K\,\log(1/\mu)\,|\cup_{k=1}^{\infty} S_k^\mu|,$$

for $0 < \mu \le \mu_0$ with μ_0 sufficiently small ($\mu_0 = \min\{1/2n, \epsilon_0\}$ where ϵ_0 is the constant in Lemma 6.5.2). The function $\psi(\mu) = 2\,K\,\log(1/\mu)$ is decreasing on $(0, \mu_0]$ and tends to $+\infty$ as $\mu \to 0$. If δ is such that $\dfrac{1}{\sqrt{\delta}} > \psi(\mu_0)$, then we can pick $\mu \in (0, \mu_0]$ such that $\dfrac{1}{\sqrt{\delta}} = \psi(\mu)$ and we obtain

$$|\mathcal{O}| \le \sqrt{\delta}\,|\cup_{k=1}^{\infty} S_k^\mu| \le \sqrt{\delta}\,|\cup_{k=1}^{\infty} S_k|.$$

This completes the proof of the covering theorem. ∎

6.6 Regularity of the convex envelope

Let Ω be a bounded convex domain in $\mathbb{R}^n$, and $u \in C(\overline{\Omega})$. Let Γ_u be the convex envelope of u in Ω defined by (1.4.1), and $\mathcal{C}$ the contact set, that is

$$\mathcal{C} = \{x \in \overline{\Omega} : u(x) = \Gamma_u(x)\}.$$

The goal in this section is to show the following.

Proposition 6.6.1 *Let $u \in C^2(\overline{\Omega})$ such that $u = 0$ on $\partial\Omega$ and $u < 0$ in Ω. Then $\Gamma_u \in C^{1,1}(\Omega)$ and $\det D^2\Gamma_u(x) = 0$ for almost all $x \in \Omega \setminus \mathcal{C}$.*

Lemma 6.6.2 *Suppose $u \in C(\overline{\Omega})$. Let $x_0 \in \Omega \setminus \mathcal{C}$ and $L(x) = \alpha + p \cdot x$ a supporting hyperplane to Γ_u at x_0. Then there exist at most $n + 1$ points $x_i \in \mathcal{C}$ such that*

$$x_0 = \sum_{i=1}^{n+1} \lambda_i \, x_i,$$

where $\lambda_i \geq 0$, $\sum_{i=1}^{n+1} \lambda_i = 1$, and $L(x_i) = \Gamma_u(x_i) = u(x_i)$, $i = 1, \ldots, n + 1$. In addition, if $u \in C^1(\overline{\Omega})$ then $p = Du(x_i)$, $i = 1, \ldots, n + 1$.

Proof. We have $\Gamma_u(x_0) < u(x_0)$. Since $L(x)$ is a supporting hyperplane to Γ_u at x_0, we have $\Gamma_u(x) \geq L(x)$ for all $x \in \Omega$ and $\Gamma_u(x_0) = L(x_0)$. Since $u(x) \geq \Gamma_u(x)$, it follows that $u(x) \geq L(x)$ for all $x \in \Omega$.

Let

$$H = \{x \in \overline{\Omega} : u(x) = L(x)\}.$$

First, $H \neq \emptyset$. Otherwise, $u(x) > L(x)$ in $\overline{\Omega}$ and by compactness $u(x) - L(x) \geq \delta > 0$ on the same set and for some $\delta > 0$. By definition of convex envelope, $\Gamma_u(x) \geq L(x) + \delta$ in $\overline{\Omega}$ and so $\Gamma_u(x_0) \geq L(x_0) + \delta = \Gamma_u(x_0) + \delta$. A contradiction.

Second, it is clear that H is closed.

Third, $H \subset \mathcal{C}$. Indeed, let $z \in H$, then $u(z) = L(z) \leq \Gamma_u(z)$. Hence $u(z) = \Gamma_u(z)$.

Fourth, $x_0 \in Con(H)$, the convex hull of H. Assume by contradiction that $x_0 \notin Con(H)$ and let N be a neighborhood of $Con(H)$ and $\ell(x)$ an affine function such that $\ell(x_0) > 0$ and $\ell(x) < 0$ in N. We have

$$\min\{u(x) - L(x) : x \in \overline{\Omega} \setminus N\} \geq \delta > 0.$$

Then there exists $\epsilon > 0$ such that $u(x) - L(x) \geq \epsilon\ell(x)$ for all $x \notin N$. On the other hand, $u(x) - L(x) \geq 0 \geq \epsilon\ell(x)$ in N. Therefore $u(x) - L(x) \geq \epsilon\ell(x)$ in Ω, i.e., $u(x) \geq L(x) + \epsilon\ell(x)$ and consequently $\Gamma_u(x) \geq L(x) + \epsilon\ell(x)$ for all $x \in \Omega$. Since $L(x_0) = \Gamma_u(x_0)$, letting $x = x_0$ we obtain $\ell(x_0) < 0$, a contradiction.

Therefore by Caratheodory's theorem [Sch93, Theorem 1.1.3, p. 3]

$$x_0 = \sum_{i=1}^{n+1} \lambda_i \, x_i,$$

where $\lambda_i \geq 0$, $\sum_{i=1}^{n+1} \lambda_i = 1$, and $x_i \in H$. Then $u(x_i) = L(x_i)$ and since $u(x_i) \geq \Gamma_u(x_i) \geq L(x_i)$, we obtain $u(x_i) \geq \Gamma_u(x_i) \geq L(x_i)$. Hence L is a supporting hyperplane to u at x_i. If $u \in C^1(\overline{\Omega})$ then

$$L(x) \leq u(x) = u(x_i) + Du(x_i) \cdot (x - x_i) + o(|x - x_i|), \qquad (6.6.1)$$

as $x \to x_i$. We have $L(x) = u(x_i) + p \cdot (x - x_i)$ and therefore inserting this expression in (6.6.1) and letting $x \to x_i$ yields $p = Du(x_i)$.   ∎

Lemma 6.6.3 *Let $u \in C^2(\overline{\Omega})$. If $x_0 \in \mathcal{C} \cap \Omega$, then there exist constants $\delta > 0$, $M > 0$, depending only on u (bounded by the C^2-norm of u in $\overline{\Omega}$) such that*

$$\Gamma_u(x) \leq \Gamma_u(x_0) + p \cdot (x - x_0) + M \left(|x - x_0|^2 \right),$$

for all $x \in B_\delta(x_0) \cap \Omega$, with $p = Du(x_0)$.

Proof.   By the Taylor expansion

$$u(x) = u(x_0) + Du(x_0) \cdot (x - x_0) + \frac{1}{2} \langle D^2 u(x_0)(x - x_0), x - x_0 \rangle + o(|x - x_0|^2),$$

as $x \to x_0$. Since $\Gamma_u(x) \leq u(x)$ and $\Gamma_u(x_0) = u(x_0)$, it follows that

$$\Gamma_u(x) \leq \Gamma_u(x_0) + Du(x_0) \cdot (x - x_0) + \frac{1}{2} \langle D^2 u(x_0, t_0)(x - x_0), x - x_0 \rangle + \epsilon (|x - x_0|^2),$$

for some $\epsilon > 0$ and $|x - x_0| < \delta$. The lemma then follows with $M = \| D^2 u(x_0) \| + \epsilon$.   ∎

Proof.   (of Proposition 6.6.1) We now assume that $u = 0$ on $\partial\Omega$, and $u < 0$ in Ω. Suppose $x_0 \in \Omega \setminus \mathcal{C}$, and let K be compact, $K \subset \Omega$, $x_0 \in K$, and L a supporting hyperplane to Γ_u at x_0, and x_i the corresponding points from Lemma 6.6.3, $x_0 = \sum_{i=1}^{n+1} \lambda_i x_i$. We claim that there exist a compact $K_0 \subset \Omega$, a constant $C > 0$, and $1 \leq j \leq n + 1$, depending only on K and u, such that $\lambda_j > C$ and $x_j \in K_0$. Indeed, let $-\delta_0 = \max_K u < 0$. Since $u = 0$ in $\partial\Omega$, $u < 0$ in Ω and $u \in C(\overline{\Omega})$, there exists a compact set $K_0 \subset \Omega$ such that $u > -\dfrac{\delta_0}{n+1}$ in $\overline{\Omega} \setminus K_0$. We have $L(x_0) = \sum_{i=1}^{n+1} \lambda_i L(x_i) = \sum_{i=1}^{n+1} \lambda_i u(x_i)$, and $-\delta_0 \geq u(x_0) \geq L(x_0) = \sum_{i=1}^{n+1} \lambda_i u(x_i)$. Hence $\delta_0 \leq (n + 1) \max_{1 \leq i \leq n+1} \{\lambda_i(-u(x_i))\}$. Relabeling the indices, we may assume that the maximum is attained when $i = 1$. Then $\delta_0 \leq$

$(n+1)\lambda_1(-u(x_1))$. Since $\lambda_1 \leq 1$, we get $u(x_1) \leq -\dfrac{\delta_0}{n+1}$ and consequently,

$x_1 \in K_0$. We also get $\lambda_1 \geq \dfrac{\delta_0}{(n+1)(-u(x_1))}$, and since $u(x_1) \geq \min_\Omega u$, we obtain

$$\lambda_1 \geq \frac{\delta_0}{(n+1)(-\min_\Omega u)} = C.$$

Recall that $x_0 \in \Omega \backslash \mathcal{C}$ and $x_0 = \sum_{i=1}^{n+1} \lambda_i x_i$, with $\lambda_1 > C$ and $x_1 \in \Omega \cap \mathcal{C}$; $L(x)$ is a supporting hyperplane to Γ_u at x_0 and by Lemma 6.6.2, $L(x_i) = \Gamma_u(x_i) = u(x_i)$, $L(x) = u(x_i) + Du(x_i) \cdot (x - x_i)$, and L is a supporting hyperplane to Γ_u at x_i, $i = 1, \ldots, n+1$.

Let $h < \mathrm{dist}(K, \partial\Omega)$. We write

$L(x_0 + h) \leq \Gamma_u(x_0 + h)$

$$= \Gamma_u\left(\sum_{i>1} \lambda_i x_i + \lambda_1\left(x_1 + \frac{h}{\lambda_1}\right)\right)$$

$$\leq \sum_{i>1} \lambda_i \Gamma_u(x_i) + \lambda_1 \Gamma_u\left(x_1 + \frac{h}{\lambda_1}\right)$$

$$\leq \sum_{i>1} \lambda_i L(x_i) + \lambda_1\left(L\left(x_1 + \frac{h}{\lambda_1}\right) + M\left|\frac{h}{\lambda_1}\right|^2\right), \quad \text{by Lemma 6.6.3}$$

$$= L\left(\sum_{i=1}^{n+1} \lambda_i x_i + h\right) + \frac{M}{\lambda_1}|h|^2$$

$$= L(x_0 + h) + \frac{M}{\lambda_1}|h|^2.$$

We claim that $\Gamma_u(x)$ is affine in the simplex generated by $\{x_i\}_{i=1}^{n+1}$.

In fact, let $x = \sum \mu_i x_i$ with $\mu_i \geq 0$ and $\sum \mu_i = 1$. Since $\Gamma_u(x_i) = L(x_i)$ and $\Gamma_u(x) \geq L(x)$ for all x, we get

$$L(x) \leq \Gamma_u(\sum \mu_i x_i) \leq \sum \mu_i \Gamma_u(x_i)$$
$$\leq \sum \mu_i \Gamma_u(x_i) = \sum \mu_i L(x_i) = L(x),$$

and so $\Gamma_u(\sum \mu_i x_i) = L(\sum \mu_i x_i)$ which proves the claim.

Consequently, $\det D^2\Gamma_u(x) = 0$ for x in the simplex generated by $\{x_i\}_{i=1}^{n+1}$ and in particular for $x = x_0$. ∎

6.7 Notes

The $W^{2,p}$-estimates for the Monge–Ampère equation in this chapter are from [Caf90a]. See also related results by Urbas [Urb88]. For examples of weak solutions u to Monge–Ampère equations with f continuous and positive such that

$D^2 u \notin L^\infty$ see [Wan95, p. 845] and [Wan92]. Lemma 6.5.2 is taken from [CG96] and it can be extended to metric balls, for example, in the Heisenberg group [Gut]. The estimates of this chapter have been recently extended to the parabolic Monge–Ampère operator $-u_t \det D_x^2 u$ in [GH].

Bibliography

[AFT98] H. Aimar, L. Forzani, and R. Toledano. Balls and quasi-metrics: a space of homogeneous type modelling the real analysis related to the Monge–Ampère equation. *J. Fourier Anal. Appl.*, 4(4-5):377–381, 1998.

[Ale61] A. D. Aleksandrov. Certain estimates for the Dirichlet problem. *Soviet Math. Dokl.*, 1:1151–1154, 1961.

[Ale68] A. D. Aleksandrov. Majorization of solutions of second-order linear equations. *Amer. Math. Soc. Transl.*, 2(68):120–143, 1968.

[Bak61] I. J. Bakelman. On the theory of quasilinear elliptic equations. *Sibirsk. Mat. Ž.*, 2:179–186, 1961.

[Bak94] I. J. Bakelman. *Convex analysis and nonlinear geometric elliptic equations*. Springer-Verlag, Berlin, 1994.

[BF87] T. Bonnesen and W. Fenchel. *Theory of Convex Bodies*. BCS Associates, Moscow, Idaho, USA, 1987.

[Cab97] X. Cabré. Nondivergent elliptic equations on manifolds with nonnegative curvature. *Comm. Pure Appl. Math.*, 50(7):623–665, 1997.

[Caf89] L. A. Caffarelli. Interior a priori estimates for solutions of fully nonlinear elliptic equations. *Ann. of Math.*, 130:189–213, 1989.

[Caf90a] L. A. Caffarelli. Interior $W^{2,p}$ estimates for solutions of the Monge–Ampère equation. *Ann. of Math.*, 131:135–150, 1990.

[Caf90b] L. A. Caffarelli. A localization property of viscosity solutions to the Monge–Ampère equation and their strict convexity. *Ann. of Math.*, 131:129–134, 1990.

[Caf91] L. A. Caffarelli. Some regularity properties of solutions of Monge–Ampère equation. *Comm. Pure Appl. Math.*, 44:965–969, 1991.

[Caf92] L. A. Caffarelli. Boundary regularity of maps with convex potentials. *Comm. Pure Appl. Math.*, 45:1141–1151, 1992.

[Caf93]   L. A. Caffarelli. A note on the degeneracy of convex solutions to the Monge–Ampère equation. *Comm. Partial Differential Equations*, 18(7&8):1213–1217, 1993.

[Caf96]   L. A. Caffarelli. Monge–Ampère equation, div-curl theorems in Lagrangian coordinates, compression and rotation. Lecture notes, 1996.

[Cal58]   E. Calabi. Improper affine hyperspheres of convex type and a generalization of a theorem by K. Jörgens. *Michigan Math. J.*, 5(2):105–126, 1958.

[CC95]   L. A. Caffarelli and X. Cabré. *Fully nonlinear elliptic equations*, volume 43 of *American Mathematical Society Colloquium Publications*. American Mathematical Society, Providence, RI, 1995.

[CG96]   L. A. Caffarelli and C. E. Gutiérrez. Real analysis related to the Monge–Ampère equation. *Trans. A. M. S.*, 348(3):1075–1092, 1996.

[CG97]   L. A. Caffarelli and C. E. Gutiérrez. Properties of the solutions of the linearized Monge–Ampère equation. *Amer. J. Math.*, 119(2):423–465, 1997.

[CH53]   R. Courant and D. Hilbert. *Methods of Mathematical Physics*. John Wiley & Sons, New York, 1953.

[CIL92]   M. G. Crandall, H. Ishii, and P-L. Lions. User's guide to viscosity solutions of second order partial differential equations. *Bull. Amer. Math. Soc. (N.S.)*, 27(1):1–67, 1992.

[CNS84]   L. A. Caffarelli, L. Nirenberg, and J. Spruck. The Dirichlet problem for nonlinear second-order elliptic equations I. Monge–Ampère equation. *Comm. Pure Appl. Math.*, 37:369–402, 1984.

[CY77]   S-Y. Cheng and S-T. Yau. On regularity of the Monge–Ampère equation $\det(\dfrac{\partial^2 u}{\partial x_i \partial x_j}) = f(x, u)$. *Comm. Pure Appl. Math.*, XXX:41–68, 1977.

[CY86]   S-Y. Cheng and S-T. Yau. Complete affine hypersurfaces, Part I. The completeness of affine metrics. *Comm. Pure Appl. Math.*, XXXIX:839–866, 1986.

[dG75]   M. de Guzmán. *Differentiation of integrals in R^n*, volume 481 of *Lecture Notes in Mathematics*. Springer-Verlag, Berlin-New York, 1975.

[EG92]   L. C. Evans and R. F. Gariepy. *Measure Theory and Fine Properties of Functions*. CRC Press, Boca Raton, FL, 1992.

[GH]   C. E. Gutiérrez and Qingbo Huang. $W^{2,p}$-estimates for the parabolic Monge–Ampère equation. Arch. Rational Mech. Anal., to appear.

[GH98]   C. E. Gutiérrez and Qingbo Huang. A generalization of a theorem by Calabi to the parabolic Monge–Ampère equation. *Indiana Univ. Math. J.*, 47(4):1459–1480, 1998.

[GH00]   C. E. Gutiérrez and Qingbo Huang. Geometric properties of the sections of solutions to the Monge–Ampère equation. *Trans. A. M. S.*, 352:4381–4396, 2000.

[GT83]   D. Gilbarg and N. S. Trudinger. *Elliptic Partial Differential Equations of Second Order*. Springer-Verlag, New York, 1983.

[Gut]   C. E. Gutiérrez. A covering lemma of Besicovitch type on the Heisenberg group $\mathbb{H}^n$. Preprint.

[Hua99] Qingbo Huang. Harnack inequality for the linearized parabolic Monge–Ampère equation. *Trans. A. M. S.*, 351:2025–2054, 1999.

[Jer91] D. Jerison. Prescribing harmonic measure on convex domains. *Invent. Math.*, 105:375–400, 1991.

[Joh48] F. John. Extremum problems with inequalities as subsidiary conditions. *Courant Anniversary Vol.*, pages 187–204, 1948.

[Jör54] K. Jörgens. Über die lösungen der differentialgleichung $rt - s^2 = 1$. *Math. Ann.*, 127:130–134, 1954.

[KS80]   N. V. Krylov and M. V. Safonov. A property of the solutions of parabolic equations with measurable coefficients. *Izv. Akad. Nauk SSSR Ser. Mat.*, 44(1):161–175,239, 1980.

[Mil97] J. W. Milnor. *Topology from the differentiable viewpoint*. Princeton Landmarks in Mathematics. Princeton U. Press, Princeton, NJ, 1997.

[Nit57] J. C. C. Nitsche. Elementary proof of Bernstein's theorem on minimal surfaces. *Ann. of Math.*, 66(3):543–544, 1957.

[Pog64] A. V. Pogorelov. *Monge–Ampère equations of elliptic type*. P. Noordhoff, Ltd., Groningen, Netherlands, 1964.

[Pog71] A. V. Pogorelov. On the regularity of generalized solutions of the equation $\det(\dfrac{\partial^2 u}{\partial x^i \partial x^j}) = \phi(x^1, \ldots, x^n) > 0$. *Soviet Math. Dokl.*, 12(5):1436–1440, 1971.

[Pog72] A. V. Pogorelov. On the improper convex affine hyperspheres. *Geometriae Dedicata*, 1:33–46, 1972.

[Pog73] A. V. Pogorelov. *Extrinsic geometry of convex surfaces*, volume 35 of *Translations of Math. Monographs*. A. M. S., Providence, RI, 1973.

[Pog78] A. V. Pogorelov. *The Minkowski Multidimensional Problem*. John Wiley & Sons, Washington, D. C., 1978.

[Puc66] C. Pucci. Limitazioni per soluzioni di equazioni ellittiche. *Ann. Mat. Pura Appl.*, 4(74):15–30, 1966.

[RT77] J. Rauch and B. A. Taylor. The Dirichlet problem for the multidimensional Monge–Ampère equation. *Rocky Mountain J. Math.*, 7(2):345–364, 1977.

[Sch93] R. Schneider. *Convex bodies: the Brunn–Minkowski theory*, volume 44 of *Encyclopedia of Math. and its Appl.* Cambridge U. Press, Cambridge, UK, 1993.

[Ste93] E. M. Stein. *Harmonic Analysis: Real-Variable Methods, Orthogonality, and Oscillatory Integrals*, volume 43 of *Princeton Math. Series*. Princeton U. Press, Princeton, NJ, 1993.

[Urb88] J. I. E. Urbas. Regularity of generalized solutions of Monge–Ampère equations. *Math. Z.*, 197:365–393, 1988.

[Wan92] Xu-Jia Wang. Remarks on the regularity of Monge–Ampère equations. In *Proceedings of the international conference on nonlinear pde, Hangzhou, 1992*. Academic Press, Beijing, 1992.

[Wan95] Xu-Jia Wang. Some counterexamples to the regularity of Monge–Ampère equations. *Proc. Amer. Math. Soc.*, 123(3):841–845, 1995.

Index

Progress in Nonlinear Differential Equations and Their Applications

Editor
Haim Brezis
Département de Mathématiques
Université P. et M. Curie
4, Place Jussieu
75252 Paris Cedex 05
France
and
Department of Mathematics
Rutgers University
New Brunswick, NJ 08903
U.S.A.

Progress in Nonlinear Differential Equations and Their Applications is a book series that lies at the interface of pure and applied mathematics. Many differential equations are motivated by problems arising in such diversified fields as Mechanics, Physics, Differential Geometry, Engineering, Control Theory, Biology, and Economics. This series is open to both the theoretical and applied aspects, hopefully stimulating a fruitful interaction between the two sides. It will publish monographs, polished notes arising from lectures and seminars, graduate level texts, and proceedings of focused and refereed conferences.

We encourage preparation of manuscripts in some form of TeX for delivery in camera-ready copy, which leads to rapid publication, or in electronic form for interfacing with laser printers or typesetters.

Proposals should be sent directly to the editor or to: Birkhäuser Boston, 675 Massachusetts Avenue, Cambridge, MA 02139